KING OF FISH

KING OF FISH

The Thousand-Year Run of Salmon

DAVID R. MONTGOMERY

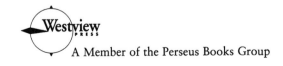

A Member of the Perseus Books Group

Copyright © 2003 by Westview Press, A Member of the Perseus Books Group.

Published in the United States of America by Westview Press, A Member of the Perseus Books Group, 5500 Central Avenue, Boulder, Colorado 80301–2877, and in the United Kingdom by Westview Press, 12 Hid's Copse Road, Cumnor Hill, Oxford OX2 9JJ.

Find us on the World Wide Web at www.westviewpress.com.

Westview Press books are available at special discounts for bulk purchases in the United States by corporations, institutions, and other organizations. For more information, please contact the Special Markets Department at the Perseus Books Group, 11 Cambridge Center, Cambridge, MA 02142, or call (617) 252–5298, (800) 255–1514 or e-mail j.mccrary@perseusbooks.com.

Library of Congress Cataloging-in-Publication Data
Montgomery, David R., 1961–
    King of fish : the thousand-year run of salmon / David R. Montgomery.
        p.   cm.
    Includes bibliographical references and index. (p.   ).
    ISBN 0-8133-4147-7
    1. Salmon.   2. Endangered species.   3. Salmon—Effect of habitat modification on.
4. Fishery conservation—Northwest, Pacific.   I. Title.
QL638.S2M646   2003
333.95'656'09795—dc21
                                                                                2003014603

The paper used in this publication meets the requirements of the American National Standard for Permanence of Paper for Printed Library Materials Z39.48–1984.
Typeface used in this text:

10   9   8   7   6   5   4   3   2   1

CONTENTS

PREFACE

SALMON. THE WORD CONJURES UP IMAGES OF THE PACIFIC Northwest. But not too long ago salmon also filled the rivers of New England and before that Great Britain. What happened?

The stories of declining salmon runs are remarkably parallel across the English-speaking world, yet the similarities are not well known even by people running salmon-recovery efforts. I trained to study how rivers and landslides shape topography and only came to study salmon after moving to Seattle to take a faculty position in geology at the University of Washington, where it is hard to study rivers without thinking about salmon. Intrigued by the connections between rivers and salmon, I began to explore how the changing landscapes of the Pacific Northwest affected the evolution and abundance of salmon. Digging into the history of the Atlantic and Pacific salmon fisheries, I found a fascinating story of how valuable public resources can gradually decline despite high-profile concerns over conservation.

Izaac Walton dubbed salmon the King of Fish in 1653. Today we know that much of what Walton wrote about salmon in *The Compleat Angler* is wrong. Yet, ironically, as knowledge of the salmon, their amazing life history, and their basic habitat requirements grew, the human impacts on salmon and their environment accelerated even faster. We now know more about the natural history of salmon than about how to save them.

Scientists typically argue that we need to do more research, that our understanding of complex natural systems is not yet complete enough, to confidently forecast their behavior. In a general sense this certainly is true. Much remains to be learned about salmon and rivers. The need for more research comes up all the time in salmon-recovery efforts. I've made this argument myself at times. But as I learned more about both the history of salmon fisheries and the effects of habitat change on rivers and salmon, I realized that knowledge alone is not enough.

Salmon are not in trouble because people didn't know about the impacts of human actions on salmon runs. Forty years ago, in his opening speech to the Second Governor's Conference on Pacific Salmon in Seattle in January 1963, Washington state governor Albert D. Rosellini declared, "We are presently faced with a desperate situation on salmon. . . .[T]he ugly truth is that if we continue as we have during the past few years, our salmon stocks are doomed to extinction!" Ignorance was not the primary problem; neither was an incomplete knowledge of the natural history of salmon. The King of Fish is not in trouble because people don't care about salmon. Laws to protect salmon have been on the books for over a century in the Pacific Northwest, and attempts to save salmon date back hundreds of years in England. Efforts to save the Columbia River salmon began well before dams tamed the river. The biggest problem for salmon lies elsewhere—in the way we make decisions and in the mismatched time scales over which societal and evolutionary processes operate, as well as the slow accumulation of little changes into large impacts that can radically alter natural systems. Under human influences the landscape gradually evolved right out from under salmon.

This book is not meant to be a rallying cry to save wild salmon at all cost in every stream across the Pacific Northwest. Neither is it intended to support the rationalizations of those who seek to write wild salmon off in favor of unfettered land use. I don't care more about fish than I do about people. Neither do I believe that we can bring salmon back to their former glory across the region as the human population doubles over the coming century. I suspect that advocates on both sides of the salmon wars will find things to praise and to criticize in this book. Such is the risk of writing about an emotionally and politically charged subject.

Many writers over the past century and a half have remarked that salmon and civilization appear to be mutually exclusive—that the development of the landscape for the use of modern societies must inevitably banish salmon to shrinking refuges uninhabited by people. I reject this argument. Although past experience certainly endorses this view, it is based on the faulty premise that we lack the ability to adapt our behavior to accommodate salmon. Salmon and civilization can coexist, if we so choose. I hope that this book brings some longer-term perspective to current debates over how to accommodate salmon in the changing landscape of the Pacific Northwest, where the next several decades will be pivotal in determining whether salmon survive in significant numbers. It simply would be tragic to lose wild salmon in the Pacific Northwest because we failed to learn the lessons of Scotland, England, and the Northeast. Moreover, those lessons tell us as much (or more) about our societies and ourselves as they do about salmon.

ACKNOWLEDGMENTS

I NEVER WOULD HAVE WRITTEN THIS BOOK WITHOUT THE help of many people. First and foremost, my wife, Anne Biklé, deserves special thanks for her valuable insight and advice on draft chapters, as well as her support, patience, and fine cooking. Xena the dog kept me company on writing retreats, and led me outdoors to stalk squirrels when I needed to recover misplaced perspective. Karl Yambert provided invaluable advice in editing the draft manuscript, and Holly Hodder both motivated me to start and encouraged me to persevere. Ray Troll graciously provided all of the original artwork for the book.

I am particularly indebted to several people whose work I have relied upon as sources. Foremost among those sources are R. W. Dunfield's *The Atlantic Salmon in the History of North America*, Anthony Netboy's comprehensive books on salmon (*Salmon of the Pacific Northwest: Fish vs. Dams*; *The Atlantic Salmon: A Vanishing Species?*; *The Salmon: Their Fight for Survival*), Richard Buck's *Silver Swimmer*, and *The Northwest Salmon Crisis: A Documentary History*, a collection of excerpts from historical documents edited by Joseph Cone and Sandy Ridlington (details on these and other titles mentioned here can be found in "Sources"). I cannot pretend to have been able to cover all aspects of the evolution and exploitation of salmon, and I recommend these sources, as well as Jim Lichatowich's *Salmon Without Rivers* and

Joseph Taylor's *Making Salmon* to readers interested in gaining more detailed insights into the history of salmon fisheries.

In addition, Tom Quinn's understanding of salmon ecology proved inspirational and educational on many occasions and David Bella's thoughts on the cultures and practice of engineering and science were enlightening. Tracy Potter helped with background research, saving me many a trip to the library, and Charles Kiblinger helped prepare graphics. I am also grateful to the superb staff of the University of Washington Libraries for invaluable help provided on many occasions. David and Toby Montgomery, Esther Bartfeld, Roger Wynne, Brian Collins, and George Pess graciously read and commented on various drafts. Patti Goldman and Dale Croes helped clarify some key points. I also am grateful to Joshua Roering for checking bibliographic details on materials only available at the University of Oregon library, and to Jeff Cramer of the Thoreau Institute of Lincoln, Massachusetts, for helping to check citations on short notice. Finally, I also owe a huge thank-you to John Buffington and the University of Idaho's Department of Civil Engineering for support and hospitality while I escaped from my office on a sabbatical to write. And in the end, I am indebted to exceptional colleagues and students for teaching me more about the habits and habitats of salmon than I ever imagined knowing.

KING OF FISH

HISTORY, THE FIFTH H

We all thought . . . that the supply of fish would continue forever.

Ezra Meeker, *Seventy Years of Progress in Washington,* 1921

M Y PROFESSORS WOULD NEVER HAVE IMAGINED THAT I, who had been trained as a geologist, would study salmon. Neither, for that matter, did I. Geologists learn to read evidence of changes that happen so slowly that they cannot be perceived in a lifetime. Perhaps this is why most people consider geology to concern deep time, dinosaurs and drifting continents—a science with little relevance to contemporary policy issues beyond natural hazards like earthquakes, landslides, and volcanic eruptions. But just as geologists read the rocks, they also read topography and glimpse the shadows of lost landscapes. As a geomorphologist—that variety of geologist who studies landscape evolution—I study how flowing water and the sediment it carries shape topography. In coming to understand the forces shaping the rivers and mountains of the Pacific Northwest, I learned to see the evolution and near extinction of salmon as a story of changing landscapes.

A lot of books have been written about salmon. Why write another one? Salmon are trapped between human population growth, economic development, degradation of environmental quality, and the politics of public policy. The King of Fish, whose slippery hordes once filled rivers across Europe and North America, is becoming rare, either vanished or disappearing across much of its ancestral range.

Though the fate of salmon rests in human hands, it is not clear that we will be able to save them even if our society wants to. Part of the problem lies in conflict between the inherent uncertainty of the natural sciences and the certainty demanded by policy makers when balancing natural resource protection against economic opportunities. But perhaps the biggest problem lies in the way that individual decisions accumulate into big effects: how land use gradually changed river basins into regions inhospitable to salmon over time spans far longer than social and political processes last.

Salmon evolved with and adapted to a changing, dynamic landscape that shaped their behavior and life history. By intentionally and unintentionally altering how landscapes work, modern human societies transformed whole regions into new worlds to which salmon are not well adapted. As odd as it may sound, geology—the study of changing, dynamic landscapes—provides the context for the rise and fall of salmon as the King of Fish.

Pacific salmon diverged from their cousins, the Atlantic salmon, when western North America began to crack up and the western coastal mountains rose during a geological reorganization millions of years before humans evolved. The topography and the salmon of the Pacific Northwest share a common history. Salmon evolved with the region. They are as much a part of the landscape as they are a symbol of the Northwest's natural splendor.

The fate of salmon is closely tied to changes on the land. The fall of the Pacific salmon is the direct result of both overfishing and other actions that subsequently reshaped the landscape. The story is not simple. But the basic connections are clear.

Humans have conducted at least three full-scale experiments on how well salmon adapt to a changing landscape. Salmon failed each time, first in Great Britain, then New England, and now in the Pacific Northwest. The current salmon crisis is nothing new. We kind of know how it generally works, even if the details change and we insist on ignoring (or having to relearn) old lessons. The strikingly similar history of salmon across these three regions carries clear implications for modern salmon recovery efforts.

Many changes to the landscape that contributed to the decimation of salmon populations occurred gradually, without particular thought to their demise. Some of these changes happened rapidly and for very specific purposes other than killing fish. Nonetheless, human indifference to the needs of the fish played a leading role. In medieval times, people knew that salmon disappear from rivers when dams block them from their spawning grounds. Some five hundred years later, the Oregon Territorial Constitution and early laws in Washington State required provisions for fish passage around all dams. Though well intended, such laws were not aggressively enforced. As development and changes swept across the land, modern fisheries managers placed their faith in the gamble that hatcheries could substitute for lost habitat, even as experience demonstrated otherwise. Today, when concern flares over the status of wild salmon, politics still impedes accommodating the salmon's basic needs, and hatcheries remain the backbone of plans to sustain many salmon runs.

Many share the blame for the decline of salmon in the Pacific Northwest. Not surprisingly, there is no shortage of finger-pointing: Land developers blame the fishing industry. Fishermen blame the timber industry. Loggers blame land developers. Everybody except dry-land farmers blames the dams. Some even blame hungry sea lions and fish-eating birds. And there is a long history of blaming declining salmon populations on Native American fishing. Yet even though there is a broad consensus among scientists regarding the primary factors driving salmon declines, actions to stem known causes remain either mired in institutional, corporate, and societal denial, dissipated by spin-doctoring, or thwarted by political agendas and bureaucratic inertia.

A Columbia River chinook salmon.

Policies that changed the landscape, and especially those that affected salmon most directly, resulted from both explicit decisions and implicit choices. Other choices were made by default through deferred decisions and avoidance of issues. More than a century ago, the salmon's current predicament was forecast, but warnings made little difference as changes sweeping across the land drove salmon from river after river. Although policies that reshaped the land were not intended to destroy salmon, the precarious state of salmon in the Pacific Northwest comes as no surprise.

Nevertheless, the listing since 1991 of sixteen Pacific salmon runs under the Endangered Species Act (ESA) created a regional political and policy crisis. Although the crisis could be considered a series of local issues that endanger fishing communities and traditional ways of life, the problem directly threatens a cultural icon of the Pacific Northwest and its natural wealth. The situation also represents a national crisis that compels us to address endangered species listings for charismatic emblems clinging tenuously to life across other rapidly urbanizing regions. And broadening our view even more, the fall of salmon populations is part of a global crisis as well, for how it unfolds may foretell the environmental future of many other regions and ecosystems. What does it say for the long-term prospects of endangered species around the world if one of the most prosperous regions of the richest country on Earth cannot accommodate its own icon species?

Hand-wringing and finger-pointing over the decline of salmon obscure recognition that the current crisis was predicted long ago. Our modern salmon crisis is a strikingly faithful retelling of the fall of Atlantic salmon in Europe, and again later in eastern North America. Just as in a Shakespeare festival, there are few new surprises in this show. The protagonists and forces that structure their interactions have been known for centuries. Even though we have the script in hand, the modern drama remains fraught with problems that cannot be fixed overnight and that impede efforts to rewrite the final act.

Factors influencing salmon abundance are often generalized into four H's: harvest, hydropower (dams), habitat, and hatcheries. Often overlooked is a fifth H: history. Learning from the past is important for public policy, particularly if policies have objectives such as the protection of rare and endangered species, or if policy failure irreversibly leads to extinction. Understanding the extent of historical modifications to river systems where salmon evolved can help clarify how to restore rivers and promote salmon recovery. First, we must ask how well we know how to restore our rivers. We then need to ask which rivers are in fact restorable. More difficult still may be deciding where we are willing to do what it takes to restore the salmon that live in them.

Salmon are a symbol of our time, icons of the Northwest, and an indicator of environmental quality—a river full of salmon is a healthy river. The dominant influences driving decreased salmon abundance are well known. But all too often the lessons of history and the impact of factors outside human control are ignored in salmon-recovery planning.

This book relates the rise and fall of salmon and their natural landscape, how salmon influenced and were influenced by human actions, and how this history relates to options and debate over issues at the center of salmon-recovery efforts. Some of the simplest, most logical proposals for accommodating salmon appear too radical for policy makers to even discuss, much less implement. Instead, the policy menu features actions least onerous to those allowed a place at the table. Consequently, salmon-recovery efforts can fail to deal with well-known causes of salmon declines and focus instead on placebos or failed approaches from the past.

Much of what is known about Atlantic and Pacific salmon is not widely appreciated by the general public, and elected officials wrangling over salmon recovery may appreciate it even less. Scientists working on salmon recovery also generally know little of the history that led to, frames, and constrains the current crisis. With legions of professionals engaged in salmon recovery, it remains rare to hear policy makers or anyone else acknowledge that how we live on the land leads directly (and sometimes indirectly) to the risk of local or regional salmon extinction. We seldom, if ever, hear a public official admit that the decline of salmon has been an implicit, even if inadvertent, policy for over a century. And yet unless we address the fundamental underlying issues, we may well spend a lot of money and still end up with no fish to show for it.

Saving salmon in the Pacific Northwest will not succeed as a surgical effort orchestrated by fishery technocrats. If history has a lesson here, it is that technological fixes and politically motivated half-measures will at best delay the inevitable. Both science and past experience show that restoring salmon runs will require reshaping our relationship to the landscape, guided by the humility to admit that we do not know how to manufacture, let alone manage, a natural ecosystem. It will also require recognition that we cannot simply engineer our way out of this crisis, as has been decreed so often in the past.

Will we be able to recover the salmon and remove it from the endangered species list, as we have succeeded in doing with the American national symbol, the bald eagle? Or will the range of salmon in the United States become restricted to Alaska and history books? Although my crystal ball is no clearer than anyone else's, one thing is certain: Decisions made now will shape the future of salmon not only in the Pacific Northwest but around the world.

SALMON COUNTRY

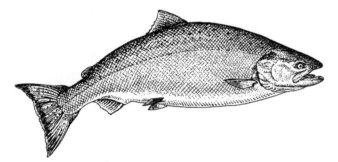

*Civilized mankind has never yet had a fresh chance of develop-
ing itself under grand and stirring influences so large as in the
Northwest.*

Theodore Winthrop, *The Canoe and Saddle*, 1863

NORMALLY I MANAGE TO AVOID SEATTLE'S INFAMOUS
traffic. Yet here I am crawling along with the morning com-
muters, braving Interstate 5 to attend the first meeting of the state's Independent Science Panel, charged with ensuring that "sound science"
is used in Washington's salmon recovery efforts. I'm not sure what I've
gotten into by agreeing to serve on the panel. But I do know that even
when seen at 70 miles an hour, the land tells a story.

This stretch of freeway crosses through a running mural of a century
and a half of landscape change. It is a four-lane time machine running
backwards from the modern Seattle-Tacoma metropolis to the less

developed Fort Lewis military reservation and the narrow strip of virtually undisturbed forest along the floodplain of the Nisqually River. Heading south along Puget Sound, the drive returns to the urbanized present at the state capital of Olympia.

In the 1850s, when Territorial Governor Isaac Stevens negotiated the treaties between the U.S. government and the area's tribes to acquire title to the Puget Sound region from its native inhabitants, it took weeks to make this journey from Seattle to Olympia through thick forests and across marshy floodplains cut by wood-clogged channels. Now only the narrow corridor along the Nisqually River preserves a landscape Stevens and his contemporaries would recognize. No doubt they would marvel at this highway and my pickup truck, much as I marvel at pioneer journals that tell of arduous journeys covering just a few miles a day.

With my field assistant, a black Lab–chow mix named Xena, I roll back through time in my Ford truck. Xena stares out the window watching the world slide by. Daydreaming too, I wonder about how this landscape has changed since my ancestors traveled west across America.

Seattle grew from the forest in just over a century. The first permanent European settlement was established in 1851 by Arthur Denny, whose name now graces a downtown boulevard. For the following half century, converting massive trees into shingles and boards provided a living for many and a fortune for the timber barons whose corporate ghosts still own much of the state. At the dawn of the twentieth century, the Yukon gold rush drove Seattle's first general boom as merchants prospered selling hardware to miners seduced north to Alaska. Expansion of the region's shipyards and aircraft production during World War II sparked the second economic boom, which was ended by Boeing's dark days, immortalized in the famous 1970s billboard requesting the last person leaving Seattle to please turn out the lights. Microsoft and biotechnology fueled the third major boom, which transformed Seattle's skyline and remodeled the verdant river valleys at the base of the Cascade Range into Silicon Valley North. Each of these periods of economic restructuring left its mark upon the landscape, and the salmon.

Seattle, Washington Territory, 1870.

I've lived in Seattle for just over ten years, a quarter of my life. Growing up in California's Silicon Valley, I watched the slow-motion paving of the fields and orchards of the Santa Clara Valley. Now, year by year, I see similar changes in Washington's landscape. The remnant old-growth forests of both private and state timberlands are gone—cut fast enough to preempt serious debate over whether they actually should be cut.

Across the state, the once impenetrable forest sports a buzz cut of angular patterns that form bizarre mosaics when viewed from commercial airliners. In a land where settlers hollowed houses out of the stumps of monstrous trees, it is now hard to find a tree big enough for a kid to hide behind. Tattered forests on the urban fringe are being cut for the second or third time to make way for suburbs and shopping malls. You have to go farther from Seattle to walk in a serious patch of old-growth forest than you do from San Francisco. Each time I drive south to Olympia, or north toward Canada, I see the concrete, pavement, and steel creep a bit farther over the Puget Lowland.

The first time that salmon compelled me to drive to Olympia was ten years ago, when I received an invitation of Phil Peterson, a biologist working with a state wildlife program. I had just arrived in Seattle, fresh from completing my doctoral thesis on where streams begin, to head a new research program at the University of Washington. The program, affectionately dubbed the "Bigdirt" project by the committee that funded it, was part of Washington's Timber, Fish and Wildlife (TFW) program, one facet of the state's efforts to defuse legal battles over declining salmon runs. The goal of the program was to understand better how humans influence landscape processes, and how salmon respond to these changes in the mountainous Pacific Northwest.

Phil was working on another TFW project, and he wanted to borrow my surveying equipment to document salmon-spawning locations on Kennedy Creek, a small stream just outside Olympia. I decided to tag along to see the fish, having never seen salmon in the wild, even though I now made my living off of them. Phil said that I wouldn't believe how many fish were in the creek.

Upon arriving at the creek, however, I was astounded not by the number of fish, but by the army of fishermen stationed along the banks. I remember wondering how any fish could make it past this human gauntlet. Only later did I learn that, unlike fishermen, salmon returning to spawn are not particularly interested in eating. This makes them sportingly difficult to catch.

In a plaid shirt and waders that made him indistinguishable from the crowd gathered around the mouth of the creek, Phil led me along the creek up a dirt road far above the row of fishing poles. We traveled down a narrow path through the woods to the valley bottom and then across the spongy floodplain. I was prepared to remain still and quietly wait for the shy creatures to grant an audience.

No need for that. I heard the fish before I saw them. And then I saw them everywhere. Splashing up the shallow stream. Hanging out in pools. Milling about, spawning, carving into the gravel streambed with their tails. Dead, rotting fish, too, draped like old shirts hung out on whatever snagged them as they floated downstream.

It is striking to see that many fish in a river. Our popular imagery celebrates the wily fish lurking in the murky depths of a dark pool—a solitary soul that can be teased out only with a well-placed fly. But this creek was crawling with the slippery beasts, bright streaks of color dancing through the stream in a chaotic ballet. These fish couldn't care less when we waded into the stream. They simply kept at it. Every now

Salmon leaping a falls on its migratory trip to spawning grounds.

and then one of them would smack into my legs as it careened purposefully yet recklessly up the creek.

For all the wild motion, for all the splashing and bolting to and fro, the salmon have a singular goal: spawning. The dance begins as the female prepares a nest in the gravel. She lies on her side and thrashes her tail, hydraulically digging a pit roughly two feet by two feet pit in the streambed. She then lays her eggs into the hole, where an eager male fertilizes them. To cover the fertilized eggs the female digs another pit just upstream. Disturbed gravel rises into the water and settles back down onto the excavated nest, burying her fertilized eggs. Once spawning is done, the exhausted fish die. Incredibly, salmon get just one shot at reproduction. If a fish's single fling doesn't produce viable progeny, its DNA is expelled from the gene pool.

The fertilized eggs develop gradually in the streambed and emerge from the gravel several months later as small fish called fry. The length of time that young salmon spend in the gravel depends on the temperature of the water. The colder the water, the longer it takes for them to develop. The total time spent in the gravel varies from species to species and place to place, as does the time of year in which spawning occurs. In western Washington, most salmon spawn in the fall, just before the rainy season. If the nest is not deep enough, the developing embryos are vulnerable to being crushed by moving gravel during winter high flows.

All those spawning fish completely rearranged Kennedy Creek. They carved away the gravel bars on the edge of the flow, evening out the streambed. Their combined efforts turned the gravel over across the entire channel, replacing stained, oxidized riverbed with truckloads of fresh, light-colored gravel.

I was impressed. That was a lot of work. As a geologist, I've seen how termites pile up huge mounds of earth in the tropics and have studied how gophers reshape hill slopes in California. I've even seen my share of valleys flooded by beaver dams. But I had never even thought that fish could rearrange streams. Yet here were the Kennedy Creek salmon changing the very structure of the stream—its width, its gravel bars, its topography. These salmon were so effective at changing their environment that it was hard to believe that their kind were endangered.

What I later came to learn amazed me more. That morning Phil and his colleagues were installing scour chains into the river. Each of these ingenious low-tech devices is a string of plastic golf balls that is anchored on one end in the streambed but whose other end floats free above. When the gravel moves during high flows, the current whisks to the end of the string any of the balls that become exposed to the flow. There they bob until someone returns to measure the depth to which the streambed has been scoured, as revealed by the number of balls at the end of the string.

These marvelously adapted fish dig nests just deeper than the depth of scour during typical winter high flows, just deep enough to allow their eggs to develop unmolested, safely buried below the dangerously mobile gravel. Some places scour more deeply than others—for example, where fallen logs constrict the current, causing it to blast out a deep hole in the streambed. So not all locations in a streambed are equally well suited for spawning. Relying on behavior honed over generations, salmon use water velocities, water depth, and gravel size as clues to good spawning locations.

The depth to which salmon bury their eggs also depends on the size of the fish. Big fish dig deep nests. Small fish dig shallow nests. The finely tuned system of digging nests to just below the depth of streambed scour during typical winter high flows means that salmon are vulnerable to both environmental changes that increase bed scour and selective fishing that takes the biggest fish. If gravel size controls scour depth, and salmon size evolved to allow them to dig deep enough to safely bury their eggs in the stable gravel beneath the streambed, then it follows that the size of streambed gravel influences the size of salmon in different rivers and streams. Large fish can spawn anywhere, since they can move gravel of any size, but small fish can only breed in small-gravel streams. Factors other than gravel size, such as habitat quality and competition with each other and other species, then take over to set population size and structure fish communities.

The fish in Kennedy Creek were chum salmon—not the most popular species at the grocery store. It is no coincidence that chum are not particularly endangered in the Pacific Northwest. That honor belongs to chinook, coho, and sockeye salmon, the tastiest salmon.

There are two genera of salmon, in the family Salmonidae. The Atlantic salmon belong to the genus *Salmo*, and the various Pacific salmon belong to *Oncorhynchus*. In addition to the five species of Pacific salmon found in the Pacific Northwest, there is an Asian species of Pacific salmon, *O. masou,* not found in North America. In contrast, the Atlantic salmon is less diverse, with variants of a single species, *Salmo salar*, distributed across Europe and eastern North America.

The scientific names of the North American species of Pacific salmon are inherited from the local Russian names reported by explorers who first described them. Each species is also known by various common names. *Oncorhynchus tschawytscha*, better known as chinook salmon, is also known as the king, tyee, or spring salmon. *Oncorhynchus kisutch*, or coho salmon, is also known as the silver salmon. *Oncorhynchus nerka*, or sockeye salmon, is also called red or blue-back salmon. *Oncorhynchus keta* is called chum or dog salmon. And *Oncorhynchus gorbuscha*, commonly called pink salmon, is also referred to as the humpback salmon or simply as the humpy. It can get confusing. From here on I'll stick with chinook, coho, sockeye, chum, and pink.

Each species reaches a different size and returns to spawn at different times. Chinook, the largest of the Pacific salmon, typically average 20 to 30 pounds and have two distinct races. Spring-run chinook enter rivers in March and April and spawn in late summer or early fall. Fall-run chinook enter rivers in September and October and spawn in fall and early winter. The largest chinook can reach an incredible 100 pounds or more. The smaller coho salmon average between 8 and 10 pounds, enter rivers and streams from September to November, and spawn in late fall or early winter. Pink salmon, which average 5 to 6 pounds, enter fresh water in August and September and then spawn in the early fall. Runs of chum salmon, which average between 10 and 12 pounds, extend from September through December but individuals typically spawn within a month after entering fresh water. Finally, sockeye salmon, which average 5 to 7 pounds, run up rivers from early June into December. The specific timing of when salmon run up rivers, and when they then spawn, varies somewhat from river to river, and sometimes between different populations of the same species in the same river.

The different species of salmon in the Pacific Northwest occupy different parts of river systems, reflecting adaptation of different life history traits and characteristics to different local conditions. Each species' predilection for certain kinds of streams is no accident. Factors like body size and run timing relate directly to the preferred habitat of each species. Chinook salmon occupy large rivers. The smaller coho salmon spawn in tributaries and are known for their jumping ability, which helps them surmount logjams and other obstructions common in small streams. Chum salmon spawn in small channels near the estuaries where freshwater meets saltwater. Pink salmon spawn in estuaries. Sockeye salmon spawn in lakes and in streams flowing into lakes. The diversification of Pacific salmon into species that live in different parts of a river system contrasts with the Atlantic salmon's more generalized use of river systems.

Although different species of salmon prefer different portions of a river, they all share a basic life history. They all begin life as fertilized eggs buried in the streambed gravel, where they incubate, hatch, and develop until they emerge as small fry. The behavior of different species of salmon begins to diverge as they emerge from the egg phase, leave the safety of the gravel, and venture out into the river. Juvenile salmon remain in freshwater for up to several years, depending on the species, and then migrate downriver to the ocean, where they spend one to four years eating and growing into the big fish that most of us recognize as salmon. Once fully grown, they return to freshwater streams and rivers, where they pair up, spawn, and then die. The unusual life cycle of migrating from freshwater to saltwater and back again is called anadromy and fish that exhibit this pattern are anadromous.

One difference between Atlantic and Pacific salmon that the fish themselves could hardly consider a detail is that whereas all Pacific salmon die after spawning, about one in ten Atlantic salmon survives spawning, returns to sea, and then returns to spawn again. Most salmon in one age class, meaning they go to sea in the same year, will come back together, and most will return to the same stream in which they hatched—but not all of them. A few come back early, and a few swim up other streams. Whether bold explorers or ineptly lost, these wayward salmon allow salmon to colonize new streams and rivers and thus are the key to the wide geographic range of the species. Occasional strays can, over time, expand salmon into new areas.

These wanderers can also recolonize streams whose original populations may have been wiped out in a disaster, for fish out at sea represent a reserve fish population for a given river. For example, if the salmon in a river are wiped out by a landslide or an erupting volcano, there are more fish from that same river waiting to mature and reenter it in coming seasons. The reserves of fish at sea are important to restocking rivers disturbed by natural catastrophes.

Salmon have several tools at their disposal to find their way back to the stream of their birth. Small amounts of the mineral magnetite secreted in their brains may function as tiny magnets that aid navigation and facilitate long migrations at sea. Some argue that in the open ocean salmon navigate by the stars, much like Polynesian navigators on transoceanic voyages. Salmon can smell differences in naturally occurring chemicals carried by runoff that impart a distinctive bouquet to a stream. Once they near the coast, smells imprinted on the journey downstream help guide the salmon back up their home river; they follow the scent of their river carried in surface waters to guide them on the final leg of their journey back from the ocean.

The natural ranges of the Atlantic and Pacific salmon are the periphery of the Pacific rim of Asia and North America, the Atlantic rim of northeastern North America, and the northern shores of Europe. Salmon are adapted to cold water and do not inhabit tropical waters. Their presence in the Southern Hemisphere is attributable to human intervention, without which they never could have crossed the warm equatorial seas. Although threatened in their Northern Hemisphere homelands, chinook salmon are thriving "down under" after being introduced to New Zealand as game fish in the early 1900s. As with many other introduced species, there is now growing concern over transplanted salmon displacing native fish. A rapidly growing salmon-farming industry in Chile and South Africa has triggered similar concerns that non-native salmon could escape into the wild, and invade southern waters. One hemisphere's endangered icon is another hemisphere's alien invader.

The awesome chinook salmon, the most endangered of the Pacific salmon in Washington State, are hard to appreciate unless one has had a direct encounter. I saw my first chinook in the early 1990s during a canoe trip down the Queets River, a day's drive west of Seattle on the Olympic Peninsula. The Queets is as close to a pristine salmon river as still exists in the lower forty-eight states. Most of the river lies within Olympic National Park, created by President Franklin D. Roosevelt in 1938 with a boundary designed to preserve at least one river in the continental United States in a natural state from its source to the sea. But Roosevelt didn't quite get the whole Queets River watershed into the Park. In the 1960s, the Washington State Department of Natural Resources blazed logging roads through the still-virgin forest outside the park boundary, ensuring that the park could not readily expand.

Logjam in the Queets River, Olympic National Park, ca. 1994.

In the late afternoon, a dorsal fin broke the surface of the Queets and headed on a collision course for our canoe. This big chinook, its back sticking awkwardly out of the water as it navigated a shallow riffle, seemed hell bent on ramming us but veered once it sensed our presence. Still speeding, it shot past our canoe toward the safety of the next deep pool, at the base of a huge natural logjam formed by fallen trees. When migrating upriver to spawn, chinook rest in pools where they can submerge and hide from predators. Once this big fish passed us and settled into the deep water we turned our attention toward not hitting the massive trees held in the next logjam.

Tim Abbe, one of my graduate students, and I were on the Queets to study how salmon-harboring pools form in a river full of old-growth logs. In particular, we were studying how the diversion of flow around large logjams scours out the deep pools favored as resting sites by migrating chinook. Our plan was to canoe down the river, measure the dimensions and depth of each pool, note the nature of the process that formed each one, and thereby determine the nature and relative importance of processes responsible for forming big pools. It turned out that in the Queets River the deepest pools form where huge trees fall into the river and snag more detritus being carried downriver. Eventually the pile becomes organized into stable logjams, which then deflect flow scouring deep holes into the riverbed. By measuring the depth of all the pools we canoed past, we found that those formed around logjams were two to four times deeper than those formed solely by the normal current of the river. More big pools means more space for more big fish, like chinook. Conversely, smaller pools provide room for fewer fish. Once they have fallen into the river, huge old-growth trees create the deep pools that provide spaces large enough to house big fish.

Stable logjams may create and enhance salmon habitat, but they are also hazards to navigation. Over the past century and a half, most Pacific Northwest rivers had logs pulled out of them and large trees cut from their banks. Consequently, the number of deep pools has decreased in rivers throughout the region. Studies of historical records and the few reaches of rivers still flowing through old-growth forest indicate that an estimated two thirds of the pools have disappeared from Puget Sound rivers over the last 150 years. Field studies also show that the abundance of chinook and coho in a river is related to the number of pools. Connect the dots and the historical loss of large pools in Puget Sound rivers significantly reduced the number of big fish that could be supported by the available habitat.

The Queets River is a historical anachronism—a large river with an intact floodplain forest. The giant trees that still fall into the Queets are not cut up for firewood or lumber. The Army Corps of Engineers does not pull them from the river. These logs remain free to do what they have done for thousands of years: sink, roll, float, pile up into huge logjams, become buried in the floodplain, and gradually decay. Observing what happens to these huge logs in river ecosystems on a canoe trip down the Queets River provides another time machine–like experience.

Knowing what rivers were like in the Pacific Northwest before development of the region helps us to assess how different our current rivers are from those in which the Pacific salmon evolved. This understanding, in turn, provides insight into how habitat changes contribute to ongoing salmon declines and how these trends might be reversed.

Intensive fieldwork on the Queets River showed that interactions between the river and the floodplain forest control pool formation, the creation and maintenance of small side channels that branch off from the main channel to rejoin it downstream, and even the topography of the valley bottom. Deep pools and quiet side channels are key elements of a good salmon stream. They form when huge logs knock sediment and water around in a river. Tim and I realized that the big trees were a critical foundation for the processes that structured and maintained the river ecosystem, and especially salmon habitat.

But how does this work? Tree trunks float, so how can they shape a river? As one might imagine, only very large logs are stable in big rivers. In addition, a root wad sticking off the end of a log can act like an anchor so that as the current pushes the log downstream it digs itself into the riverbed. Then other logs moving down the river can be pinned against the snagged log and a logjam starts to grow. A stable logjam not only diverts flow, carving out a pool where the current slams into the riverbed, but it also causes deposition of sand and gravel in the low-velocity water immediately downstream of the logjam.

Eventually, enough sediment can build up to form an island, which over time can become incorporated into the floodplain. Small backchannels left between these logjam-formed islands mean that the "river" is actually not a single channel filled with flowing water. Instead, ribbons of water moving across the valley bottom split into a complex web of small side channels where massive logjams block the main flow.

Using a combination of tree-ring counting and radiocarbon dating, we compared ages of the biggest floodplain trees to the ages of ancient logjams sticking out of the riverbanks on which the big trees grew. Most of the big trees were just a little younger than the logjams beneath their roots, demonstrating that the big trees started to grow in the sheltered zone behind a logjam. The buried logjams were themselves founded upon huge logs from an older forest stand, no doubt sheltered by an even earlier logjam. Over hundreds of years the floodplain of the Queets River was built by repetition of this process across the valley bottom.

It was startling to learn that logjams not only could prove stable over centuries but also could actually push around one of the largest rivers in the region. Most research on salmon habitat in the Pacific Northwest has focused on small streams. After all, small streams are easy to work in— you can wade across them, see the streambed, and safely take measurements without a boat. And it is not too hard to imagine that a tree falling into a stream may remain where it fell, divert the flow, and scour out a pool. Until recently, however, the perception among stream ecologists and geomorphologists was that logs just float right on down big rivers. We're still learning the extent to which logs from floodplain forests influenced rivers in the old-growth forests of the Pacific Northwest.

Following European colonization of the Pacific Northwest, big lowland rivers were the first to change, and there is little documentation or cultural memory of their prehistoric condition. Yet the first crisis involving the listing of salmon under the Endangered Species Act (ESA) in Washington State involves the chinook that live in the biggest rivers. It is an awkward truth that our scientific understanding of the processes shaping big rivers in forested regions lags behind our societal need to make decisions. Nonetheless, decisions are being made—often with inadequate consideration of what we *do* know about salmon, their habitat, and the effects of both past and potential future human actions on them.

Through all the finger-pointing and political posturing it remains clear that the interaction of geology, rivers, and fish defined the rise of the Pacific salmon, just as the interaction of people, rivers, and fish is driving their fall. Settlement of the Pacific Northwest involved massive changes to rivers and streams—and these changes have had clear, predictable impacts on salmon. Even without the benefit of hindsight, the present ESA listing of various species of Pacific salmon can be seen as the direct, logical, and foreseeable outcome of a century and a half of explicit policies and implicit choices. How we respond to this latest salmon crisis will determine whether in the future there will be salmon passed on to our children's children.

MOUNTAINS OF SALMON

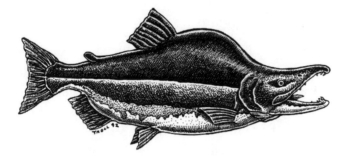

You cannot step in the same river twice, for the second time it is not the same river.

Heraclitus (535–475 B.C.)

IN 2001, ALMOST A DECADE AFTER THE TRIP TO KENNEDY Creek in the early 1990s that hooked me into studying salmon streams, Phil Peterson gave me another opportunity to see salmon in the wild, this time along the headwaters of the Skokomish River on the southeast corner of the Olympic Peninsula. Phil had left the Timber, Fish, and Wildlife program for a position as a biologist with the Simpson Timber Company. We had continued to work together on the geomorphology of salmon and in the later 1990s I had helped Phil develop ideas for a habitat conservation plan for Simpson's tree farm. Unlike our visit to Kennedy Creek years earlier, our quarry this time was fossil salmon.

A local fisherman, Jeff Heinis, had noticed large fish fossils protruding from the bank of a remote reach of the river on Simpson Timber Company land. Phil followed up on the report and went out to see the fossils. Upon returning he called me to ask if the University of Washington had a paleontologist interested in salmon, and whether I was interested in helping interpret the deposit that contained the fossils. Phil was concerned that erosion of the riverbank could destroy the fossils during winter high flows. I referred him to the Burke Museum, the University of Washington's natural history museum, and its experienced fossil collectors and curators.

He then emailed me photos of the fossils. They were remarkable. The fish were quite large, and numerous. The outcrop of sediment that held the fossils was set in the bottom of a deep canyon, and the fossils were in lake deposits within the canyon. Layer upon layer of fine-grained sediment formed a series of small "steps" or terracettes down to the river's edge. Fossil fish lay in the sediment like pressed flowers. The photos revealed relatively complete skeletons and impressions of scales and other fine features that had been preserved when the fish settled into the soft mud at the bottom of the ancient lake.

This was too good to pass up. After all, fieldwork is one of the basic pleasures of geology. And Xena the dog would be appalled if I turned down an invitation to work along a stream or river where she could explore. So we piled into my truck and headed south to meet Phil in a Denny's parking lot in Olympia. After I introduced Phil to Xena, we transferred our gear into Phil's unmarked company truck and drove over Kennedy Creek, past Simpson's main office, and out to the floodplain of the Skokomish River.

The South Fork of the Skokomish begins in steep U.S. Forest Service land deep in the Olympic Mountains. On its way to Puget Sound, the river cuts through a narrow canyon and then traverses Simpson timberlands. After leaving the mountains, the river flows across a broad floodplain and deposits its gravel load. Farms and houses now cover the floodplain.

The North Fork of the river begins in Olympic National Park but is diverted into a tunnel that flows through a ridge directly to an arm of

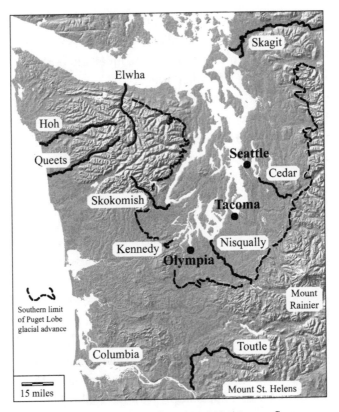

Selected rivers and creeks of western Washington State.

Puget Sound, powering turbines along the way to light the city of Tacoma. This diversion means that less water now reaches the confluence with the South Fork. Downstream, the river completes its journey through the Skokomish Indian Reservation before emptying into the sound. We drove up along the river past the canyon on the South Fork and into the headwaters.

We reached the end of the road in late morning and parked on a broad upland that formed a surface a few hundred feet above the river, where a logging track became a foot trail. Well past the end of the road we started to descend a narrow path that led toward the canyon. As usual, Xena was in the lead scouting for good smells to savor and critters to chase. I could see how glaciers had planed into a gentle upland the landscape into which the river had then carved its canyon. At the edge of the canyon the terrain dropped over a series of terraces, the lowest of which formed a narrow valley bottom, to which Phil led us.

The fossil-laden outcrop was just upstream of where the canyon narrows. Apparently ice had dammed the river at some time in the distant geological past, creating a lake that entombed the fossils in its sediments. The terraces over which we dropped to get down to the river were like geological bathtub rings that recorded the progressive incision of the river after the ice retreated back down the valley and home toward Canada.

Xena lost no time in getting down to business, scampering down the riverbank and starting to paw at fossil salmon protruding from the bank. "Leave it!" echoed off the canyon walls. In response I got the soulful-brown-eyes treatment. But this was serious business. I responded with the coolest "Yes, I really mean it" that my eyes can muster.

The fossil salmon were best preserved in the lowest levels of the lake sediments, the earliest chapters of this geological story. Immediately below the lake sediment and its beautiful salmon skeletons lay the gravel of an ancient streambed. We collected sticks and wood in the lake sediments to be radiocarbon dated, a technique that measures the amount of carbon-14 (an isotope present in all living matter that is subject to radioactive decay) that is left in the organic material being dated. It turned out that all the carbon-14 maintained by living tissue had disappeared, indicating that the samples were older than 40,000 years—the limit of conventional radiocarbon dating. So the lake formed well before the last glacier overran Puget Sound less than 20,000 years ago.

Once back at the university I started inquiring whether anyone had studied glacial lake deposits around the Skokomish River. It turned out that in the next valley over, U.S. Geological Survey scientists had found million-year-old volcanic ash interbedded, or inter-layered, in deposits almost identical to those containing the fossil salmon. It seemed our lake had been formed about a million years and several glaciations ago, in the middle of the Pleistocene Epoch, the most recent 2 million years of geologic time. We had an outcrop that contained million-year-old salmon.

Not long after my first visit to the fossil site with Phil Peterson I met Gerald Smith, a renowned expert in salmon paleontology at the University of Michigan. Over dinner we discussed the evolution of salmon and I invited him to visit the Skokomish River site. The following summer, Jerry, Phil, and I returned to the site along with Bruce Crowley and Jim Goedert, senior fossil curators for the Burke Museum. While Bruce and Jim calved off great slabs of sediment to collect whole fish for exhibits, Jerry was on his knees with his face in the dirt painstakingly picking away at the outcrop. The finely laminated clay and coarser silts formed easily discernible bands (called varves) reflecting the alternation of slow fine-particle sedimentation in summer (clays) and more rapid sedimentation of coarser particles in winter (silts). After a couple of hours, Jerry rose and filled us in. The salmon were four-year-old sockeyes. Fossils of males and females together in the silty layers indicated that the fish had been on a spawning run when they died—for sockeye, it would have been in the fall.

That evening Jerry explained why he likes to examine little bits of bone rather than whole fish preserved on an intact slab cut from the rock. He can turn fragments around to examine them from all sides. He can see the details of their shape in three dimensions and pick apart the outcrop bit by bit to see how the ancient fish bones sit within the fine layers of sediment. Fossils stuck in a big slab of rock are beautiful, and ideal for display, but to Jerry they present just a single view—more like a photograph than the real thing.

A fundamental problem in understanding the evolution of salmon is that there is almost no fossil record of these fish. Most adult salmon die in mountain streams. As mountains erode, so do the river deposits that contain fragile salmon bones. Consequently, there is little geologic record of salmon streams, let alone salmon. For many years the evolutionary history of salmon was obscure because of this simple fact—there simply are not many salmon fossils. Finds of freshwater fish fossils, like those on the Skokomish, are few and far between.

One of the biggest puzzles for salmon paleontologists has been where and how the salmon survived during the repeated glaciations, ice ages, that characterized the Pleistocene Epoch. Historically, the conventional wisdom was that salmon in the Pacific Northwest rode out the ten or more glacial advances between 2 million and 10,000 years ago in ice-free areas in the Columbia River Basin as well as rivers in Oregon and California. In this view, salmon gradually spread from these sanctuaries after the ice melted. But there are other possibilities. Salmon also could have survived glacial times in ice-free areas along the coast of Washington, Alaska, and perhaps even British Columbia. Today, salmon inhabit rivers at the foot of modern glaciers in Alaska. Sea level was hundreds of feet lower at the height of glaciation, owing to the immense volume of water trapped in polar ice caps, and much of the now-underwater continental shelf was exposed. Perhaps salmon escaped the glaciers by simply moving downstream into rivers flowing across the continental shelf. Salmon can inhabit glacially fed rivers, and they inhabit rivers at the foot of modern glaciers in Alaska. Although this new view is gaining acceptance, it is not yet settled as to which stocks spread from which refuges after the glaciers melted off.

Enter modern technology. Genetic analyses are being used to sort out salmon populations with different histories. DNA sequencing techniques also allow a researcher to use the accumulation of minuscule genetic errors—mutations that pile up over generations at a relatively constant rate—as a clock to estimate when populations diverged. Sequence differences in genes from Atlantic and Pacific salmon indicate that they began to diverge from one another about 20 million years ago. This was a time when cooling of the Arctic Ocean isolated populations in the Atlantic and Pacific Oceans. Freezing of the polar seas below the tolerance of salmon imposed a barrier that prevented the commingling of salmon stocks in different oceans.

DNA sequencing of North American and Asian Pacific salmon shows that a little more than 10 million years ago the Pacific salmon started branching into clans. By 6 million years ago the five different species of Pacific salmon found in North America had separated both from the Asian salmon (*O. masou*), which is most like the ancestral Pacific salmon, and from each other.

But why are there so many species of Pacific salmon and only one Atlantic salmon? What led to the striking difference in the evolutionary tra-

Salmon family tree. Courtesy of Ray Troll.

jectories for salmon on the East and West coasts of North America? The dearth of fossil evidence for salmon evolution has led to a variety of arguments for the diversification of Pacific salmon from the common ancestor they shared with the Atlantic salmon. Conventional explanations invoke advance and retreat of glaciers as the trigger for isolation, diversification, and behavioral modification of local Pacific salmon stocks into distinct species. But why would this only occur in western North America? Glaciers affected not only both coasts of North America but also Europe and northeastern Asia. Why would an evolutionary response to glaciations have been restricted to just the Pacific side of the salmon family?

Genetic analyses of mitochondrial DNA show that the modern species of Pacific salmon evolved before the glaciers began advancing 2 million years ago. But even without sophisticated biochemical tools, the idea that the North American species of Pacific salmon evolved during the age of ice can be dismissed. Geologists have been finding fossils of the modern species of salmon in deposits that predate the earliest glacial advances.

Accumulating evidence shows that most of the differentiation leading to the modern species occurred long before the start of glaciation. The oldest known fossil of ancestral salmon, *Eosalmo driftwoodensis*, found in British Columbia, is from roughly 40 million years ago. The fossil record shows that the Pacific and Atlantic salmon had diverged by 20 million to 10 million years ago, during the Miocene Epoch. By 6 million years ago, the Pliocene Epoch, species resembling the modern sockeye, pink, and chum salmon were present in Idaho and Oregon. Genetic sequencing and analyses of mitochondrial DNA in modern salmon also suggest similar date ranges for the divergences in the salmon family. This is good news for those trying to solve the puzzle of what led to the different species of Pacific salmon. The fossil record and genetic analyses each independently confirm that speciation of the Pacific salmon occurred long before the glaciers began their cycle of advance and retreat. So much for the glacial theory of salmon evolution.

What about anadromy as a basis for explaining salmon evolution? One would hope that a fish would have a good reason to swim thousands of miles and shift from freshwater to the ocean and back again. Why not just stay put, relax, and skip that epic odyssey to sea? One idea invokes changes in the relative availability of food in terrestrial versus marine environments to explain speciation of the Pacific salmon. Global cooling between 40 million and 20 million years ago led to decreased primary productivity in temperate streams and to increased marine productivity—and supply of food for salmon—through enhanced upwelling of deep marine water. It's plausible, therefore, that going to sea and returning to spawn in freshwater streams or rivers evolved in concert with the oceans' becoming more productive and terrestrial streams' becoming less productive as the climate cooled. Climate-driven changes in food availability made going to sea a wise move as streams and rivers became depleted of nutrients in the cooler climate. Even in today's warm interglacial climate, the streams in the Pacific Northwest remain nutrient-limited. Development of a sea-going life cycle may simply reflect that the Miocene oceans provided richer feeding grounds for growing salmon than the streams in which they hatched.

Recent studies show that salmon, gaining well over 90 percent of their body weight at sea, carry vast quantities of marine-origin nutrients when they return to their home streams to spawn and die. Their rotting bodies nourish stream-dwelling insects that, in turn, feed young salmon the following season. Juvenile salmon have also been known to nibble on the rotting carcasses of their elders. In this way, anadromous salmon provide an organic inheritance for their offspring by fertilizing their home stream. Salmon carcasses also provide a huge infusion of nutrients to aquatic and streamside communities as they decay, get scavenged, and become incorporated into other organisms. In some places, fisheries managers now dump carcasses of hatchery-produced salmon into streams to promote the growth of wild salmon.

This infusion of nutrients into streams sustained a diverse, interdependent ecosystem. Scavengers pull almost half the carcasses of coho salmon from small streams on the Olympic Peninsula. Twenty-five to 90 percent of the nitrogen in the bones and hair of grizzly bears in the Columbia River basin came from marine sources. More than 90 percent of the nitrogen in Alaskan brown bears comes from salmon. Marine nitrogen brought up rivers in salmon even finds its way into streamside trees. Up to a third of the nitrogen in valley-bottom forests swam up the river as a fish. Trees growing along salmon-bearing streams grow up to three times faster than those growing along salmon-free streams. For Sitka spruce along streams in southeast Alaska this shortens the time needed to grow a tree big enough to create a pool, should it fall into the stream, from over three hundred years to less than a century. Salmon fertilize not only their streams but the huge trees that create salmon habitat when they fall into the water.

Evolution of an anadromous life cycle and the resulting transfer of nutrients from oceans to freshwater ecosystems is an important aspect of salmon biology. Salmon could never grow as large as they do in the nutrient-poor environment of freshwater rivers and streams. There just isn't enough food.

Thus anadromy explains some aspects of salmon's evolution; it does not, however, explain the different evolutionary trajectories of the Pacific and Atlantic salmon, since both are anadromous. We must look beyond global marine cooling and consequent changes in nutrient availability to explain why the Pacific salmon separated from the Atlantic salmon and split into the species we know today.

Geographic isolation is the usual explanation ecologists offer for speciation. At first glance such an explanation might seem to account for divergence of the Pacific salmon from the Atlantic salmon. But the use by the North American species of Pacific salmon of different habitats within the same river systems points to environmental change rather than geographic isolation as the cause of their divergence into distinct species. Across their overlapping geographic ranges, pink and chum salmon generally spawn in small streams near the outlet of a watershed; sockeye spawn in lakes and lake inlets; coho spawn in small tributaries; and chinook spawn in major rivers and large tributaries. The segregation of Pacific salmon into different portions of river systems suggests that they evolved in different kinds of streams to take advantage of different ecological niches in mountain drainage basins rather than from the physical isolation of local populations.

I believe that the answer to the puzzle of salmon evolution lies in the strikingly different history of the landscapes of eastern and western North America. Evolutionary histories of the Atlantic and Pacific salmon parallel differences in the topographic evolution of North America. The topography of the eastern coast of the United States is ancient compared to that of the West Coast. The Appalachian range formed when Europe slammed into and crumpled up the Eastern Seaboard of North America before the opening of the Atlantic Ocean, about 200 million years ago. The Appalachians approached their current form during the reign of the dinosaurs, about 100 million years ago. Since then they have been slowly eroding away. Geologically, not much happened on the East Coast of North America in the last 100 million years, and the topography of the region has remained fairly constant since long before salmon evolved.

In contrast to the stability of eastern North America, the topography of western North America has changed significantly since the salmon family split into its Atlantic and Pacific clans. At roughly the time of the Pacific salmon's original divergence from the Atlantic salmon in the Miocene, regional geologic changes literally fragmented longstanding landscape features, including a broad plateau extending from the Rocky Mountains to the Pacific Ocean. About 16 million years ago this plateau began to split apart, collapsing to form the Basin and Range topography in what is now Nevada. On the Western Seaboard, the Alaska, British Columbia Coast, Cascade, and Oregon Coast ranges and the Olympic Mountains all rose significantly during the period between 20 million and 6 million years ago. This coincides with the period between the divergence of the Pacific and Atlantic salmon and the evolution of the modern species of Pacific salmon.

The salmon and the topography of western North America appear to have evolved together in response to tectonic forces that drove mountain building along the West Coast. Certainly, geology is not the whole explanation: Glaciation and climate change undoubtedly played roles too. Local isolation resulting from glacial advances likely helped differentiate discrete populations of salmon, and climate influence is apparent in the evolution of anadromy in both the Pacific and Atlantic salmon. But a primarily geological cause for the evolution of the Pacific salmon could explain the strikingly different evolutionary trajectories for salmon on either side of North America.

How could topographic change have split the ancestral Pacific salmon into five new species? Topographic changes along the coast and in inland watersheds diversified stream conditions. Most salmon faithfully return and spawn in their home stream, leading to a high degree of reproductive isolation between populations in different rivers. Take this concept one step further and apply it also to salmon that spawn at different times or in different places in the same river system. Voilà, you have one of the fundamental preconditions for evolving a new species—reproductive isolation.

A small percentage of returning spawners stray, a behavior that can help recolonize streams after major disturbances such as volcanic eruptions and glaciations. Straying also allows salmon to rapidly occupy new habitats as they become available. With some straying behavior and strongly heritable traits, an increasing variety of stream environments within the same geographic area could provide a viable impetus for speciation.

It may take millions of years for geological processes to reshape the landscape so that a new species emerges. But extinctions can occur rapidly. Some species of salmon have already gone extinct. The largest of the salmon, the so-called sabertooth salmon (*Oncorhynchus rastrosus*), is already gone. These monstrous fish grew to enormous lengths of up to 10 feet and weighed as much as 350 pounds. They thrived in the Pliocene, before the cold climate of the Pleistocene set in. Apparently they did not survive the transition to the glacial world.

Although salmon have had some evolutionary dead ends, they are in fact well equipped to take advantage of new environments. Like weeds colonizing a vacant lot, salmon produce lots of offspring that can move into new or empty niches. A spawning pair will produce thousands of eggs, and even though relatively few offspring make it back to spawn again, it is enough. In a population that was neither growing nor shrinking two adults should return on average for each pair of spawners. Tom Quinn, a prominent salmon ecologist at the University of Washington, compiled studies that examined the survival rate of salmon from one life stage to another and concluded that when not subject to fishing pressure or limited by habitat availability, four to six wild salmon return on average for every pair that spawned. This higher-than-replacement rate of return indicates that, left to themselves, salmon populations will tend to expand to fill the available habitat or until they become so numerous that competition for space or food limits their numbers.

The retreating glaciers of Alaska's Glacier Bay National Park provide the setting for an ongoing experiment in how fast salmon can colonize new streams. The ice front surveyed by Captain Cook in the late 1700s retreated more than 50 miles by the 1980s, exposing raw valley bottom and new streams on the valley sides. Streams closest to the modern ice have been exposed for decades, whereas areas closest to the original ice front have been free of ice for over two centuries. Once the streambed stabilized enough to be colonized by algae and then invertebrates, salmon began moving into the new streams. Those exposed for only a century were already full of sockeye, pink, and chum salmon.

In some circumstances salmon adapt rapidly to changes in established river systems. For example, an unintended experiment was conducted in the early 1900s when the outlet of Washington's Cedar River was redirected into Lake Washington to facilitate navigation, flood control, and development. Sockeye salmon need a river system that includes a lake, so fisheries managers transplanted sockeye into this new habitat. Some of the sockeye introduced into the Cedar River basin began spawning on beaches in Lake Washington and others spawned in the Cedar River. After just a matter of decades, the fish spawning in these different locations are becoming genetically distinct from each other.

If salmon can colonize new streams and evolve that fast, then why can't reintroduction of salmon in places where they have been wiped out eventually reestablish viable, naturally reproducing runs? For such efforts to prove successful the landscape must be able to support the salmon's basic life-cycle requirements—these can't change on demand. Salmon will not thrive if there is no gravel in which to bury their eggs, if eggs are scoured out of streambeds, if there is no cover to hide young salmon from predators, if pollutants poison the water, or if lack of oxygen stifles embryo development. Obviously, even the most successful reintroduction program would fail if salmon are thrust into a world that no longer supports the basic needs of all salmon. But the more subtle adaptations of salmon to the specific conditions of their home stream means that the survival of transplanted salmon also depends on an environment that has the critical characteristics of the rivers and streams in which those salmon evolved.

As Heraclitus noted, it is obvious that streams are dynamic environments. Floods can move a river around on its floodplain, and local processes of bank erosion can fill in or create new pools during even modest storms. It is also obvious that the Pacific salmon are resilient to the natural disturbances of rivers and streams in the Pacific Northwest. After all, salmon evolved and thrived in these channels. Yet it remains difficult to determine how to characterize and evaluate the natural disturbance regime (pattern)—for rivers change over a very wide range of time scales. The sudden disturbance of a pool drying up in a summer drought can kill a local population of juvenile salmon and a landslide can wipe out the salmon in an entire tributary, just as the long-term "disturbance" of a rising mountain range can help create new species.

According to the naturalist and author Richard Buck, one needs no special training to see how river conditions can literally shape salmon. In his 1993 book, *Silver Swimmer*, this avid fisherman and leader in salmon conservation efforts described how the conditions in different rivers favor distinct body styles among Atlantic salmon. Contrasting Canadian salmon from different rivers, Buck noted that deep, fast rivers have large, powerful fish, whereas small rivers have smaller, stocky fish. Similarly, he found that among Scottish rivers the wide slow-moving River Tweed has large bulky salmon, whereas swift highland streams host "lean and well-formed" salmon.

Buck also described differences in tail and fin size of salmon from rivers of different character, comparing the shape of salmon he caught in Norway's Laerdal River to those he caught in New Brunswick's Restigouche River. The Laerdal River salmon are "long, taut, and compact," with huge fins and a large tail well suited for running up the turbulent rapids that characterize the river. In contrast, the Restigouche River salmon are "full-bodied, broad, and chunky," well-suited for life in the "big, wide, deep, relatively slow-moving river."

Even among the Atlantic salmon, natural selection favors development of body forms best suited for the hydraulics of a salmon's home river. The Atlantic salmon have a diverse gene pool that acts as a genetic reservoir upon which evolutionary processes can act, and provides resilience to some environmental changes. Similar physical and genetic diversity in their ancestors probably enabled diversification of the Pacific salmon in response to their changing environment.

The rivers of the Pacific Northwest continued changing as the mountains rose and the glacial ice advanced, only to melt away and advance again and again. For almost two thousand years, from about 18,000 to 16,000 years ago, Puget Sound rivers were overrun and dammed by a massive wall of ice, the Puget Lobe of the Cordilleran Ice Sheet, which rose three times as high as Seattle's Space Needle. In parts of Puget Sound the ice was thick enough to depress the Earth's crust by hundreds of feet. Once the mountain of ice disappeared the land rebounded skyward. During the first 6,000 years after glaciers began melting, sea level rose hundreds of feet, gradually flooding coastal areas and lowland valleys before more or less stabilizing about 5,000 years ago. Certainly these events rearranged the landscape, but the changes occurred slowly. Over time, the total amount of freshwater salmon habitat increased as sediments shed from the rising land extended river valleys and estuaries out into Puget Sound.

The dynamic, geologically young landscape of the Pacific Northwest is not a quiet, safe place. Yet salmon thrived in prehistoric times even though large disturbances disrupted river ecosystems time and time again across the Pacific Northwest. Volcanoes of the Cascade Range grew and blew themselves apart in catastrophic eruptions. Hot volcanic mudflows, known as lahars, obliterated rivers and filled up valley bottoms. Mount Rainier shed massive lahars that swept down river valleys, burying everything in their path.

At least one of these lahars reached the present-day Elliot Bay, near downtown Seattle. During construction of a new park just upstream from the Port of Seattle along the Duwamish River, backhoes revealed that the valley bottom consisted of a massive sand layer resting atop estuary-bottom mud. This was odd. Marshes are not usually lined with nice clean sand. Even more curious was that the sand contained distinctive minerals from Mount Rainier, over 50 miles away. Brian Atwater, a geologist with the U.S. Geological Survey, knew that I had worked on similar deposits in the Philippines and thought I would enjoy seeing what they had unearthed along the Duwamish River. He was right.

The walls of Brian's trench displayed relationships that revealed these sands had been deposited by the river and had rapidly buried an ancient tidal flat, in the process preserving delicate vegetation on the entombed marsh surface. The most reasonable interpretation was that the sand represented the downstream end of a massive lahar. Brian pointed out that when the top of these sands was compared with those of similar deposits across the valley it indicated that they came from the same lahar, which not only buried the river but filled in the entire valley. He also found that old maps of the area show that the modern terrace formed by the lahar was the dry spot selected by the Denny party for its initial settlement at Seattle in 1851.

Other Cascade volcanoes produced similar catastrophic lahars. Massive lahars from Mount Baker and Glacier Peak killed off salmon in the rivers down which they sped. Most of the delta of the Skagit River was formed during a single massive lahar that extended the valley bottom miles out into Puget Sound, creating space for new river reaches as it went.

The decimation of the North Fork of the Toutle River by the 1980 eruption of Mount St. Helens dramatically revealed the awesome power of catastrophic events to reshape Pacific Northwest rivers. Just minutes after Mount St. Helens erupted, the Toutle River filled with boiling mudflows, obliterating life from a river once teeming with salmon. Amazingly, within just a few years salmon were already finding their way back up the river, demonstrating that salmon can quickly start to recolonize even a massively disturbed river. This rapid return to a devastated river demonstrates the advantage of a life history that includes spending several years at sea, providing a reserve of new colonists already programmed to repopulate decimated areas. As the Toutle River gradually cleans itself of volcanic debris it gradually is becoming once again a cold-water, gravel-bed river hospitable to salmon.

For millennia the salmon of the Pacific Northwest weathered large disturbances like volcanic eruptions because only a small portion of their range was affected by any one event, and such disturbances were not sustained for long periods of time. Salmon still at sea could recolonize a river after a disturbance that only lasted for a short time. As long as refuges remained from which repopulation could occur, the small percentage of individuals that stray from their home streams would eventually recolonize even catastrophically disturbed environments, such as those cooked by lahars or frozen beneath glaciers. When these wanderers found suitable empty habitats they spread to new streams. Straying behavior also builds resilience to catastrophic disturbance into a population. In a dynamic and often dangerous environment, the long-term stability and success of the species depended on the reserve of salmon out at sea and these wayward salmon—as long as disturbances and disasters were not too widespread and did not occur too often.

Volcanic eruptions are not the only disturbances that affect Northwest rivers. Along the region's coastline the earth's crust beneath the Pacific Ocean is being pushed under North America. Rocks scraped off the colliding plates pile up to form the coastal mountains. Once the slab of oceanic crust sinks deep enough it starts to melt and the rising magma feeds the volcanoes of the Cascade Range. About every five hundred years the entire coast from Northern California to Canada lurches toward New York all at once in an earthquake that releases more power than that locked in our nuclear arsenal. The associated ground shaking can trigger massive landslides that can dam rivers.

During past superquakes, huge amounts of sediment were introduced into rivers and streams when whole mountain sides collapsed. Massive landslides can dam rivers and block salmon as effectively as any man-made dam. Large landslides have dammed even the mighty Columbia River in the distant past. Although it is hard to know for how long they may have been locked out, salmon may have had to recolonize the interior Columbia River basin after landslide dams blocked their access to and from the sea. The resulting effects on fish may not be quite as severe as volcanic incineration, but large earthquakes do pose another source of disturbance for river ecosystems in the Pacific Northwest.

More frequent and less dramatic disturbances also affect the salmon. Big storms trigger floods that transform the river environment both during and sometimes even after the event. Where do fish go in floods when the river becomes a raging torrent that can take out a bridge? Even a large fish has no chance against such a current. So they hug the banks, burrow into the streambed, or cruise out into the shallower water spilling onto the floodplain. In case they don't make it, having several generations at sea at any one time acts as a hedge against the destructive effects of violent floods. Forest fires can also trigger large pulses of sediment into river systems that, in turn, can change the depth to which streambeds scour and fill, which can crush or entomb salmon embryos buried in the gravel. Low summer flows during droughts can wipe out entire classes of juvenile salmon as they become stranded in pools that dry up. The strategy of spending three to five years at sea not only provides salmon with access to more food than available in their home stream, but also buffers salmon populations against the perils of life in the Pacific Northwest's dynamic rivers and streams because at any one time, multiple generations are at sea.

Strangely enough, disturbance events that can decimate one generation of fish also over the long run help create the best habitat for future generations. Side channels that provide safe refuges for salmon during flooding of the main channel are themselves formed during floods. A large tree trunk falling into a river can cause local scour that excavates salmon embryos developing within the gravel—a negative—but the new pool also provides excellent habitat for the next years' juvenile salmon. The dynamic nature of their environment should lead us to expect salmon populations to exhibit substantial year-to-year variability.

This brings up a troubling question. If salmon are resilient enough to withstand extreme events like massive landslides, volcanic mudflows, and glaciations, then why are they going extinct across much of their range today? Recent changes in the landscape must be rendering rivers unable to sustain them. The fossil salmon of the Skokomish River are ghost icons, reminders that in just over a century, humans managed to do what repeated onslaughts of ice, in places half a mile high, could not accomplish. Will salmon repeat their ice age comeback after the human age? To recolonize a river buried by ice, degraded by human actions, or depopulated through natural disturbances, salmon must survive in protected refuges from which to spread during more favorable times.

It is sobering to think that salmon could take the worst nature could throw at them for millions of years—from floods to volcanic eruptions—but that little more than a century of exposure to the side effects of Western civilization could drive them to the edge of extinction. Humans and salmon survived together for thousands of years on both coasts of North America. Was their coexistence possible simply because there were fewer people in the region—or were Native American cultures adapted to sustain salmon fisheries?

SALMON PEOPLE

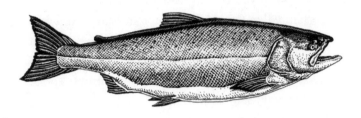

*If the salmon could speak, he would ask us to help him survive.
This is something we must tackle together.*

Bill Frank, Jr., Chairman, Northwest Indian
Fisheries Commission, 1991

A S THE U.S. CORPS OF DISCOVERY, LED BY MERIWETHER
Lewis and William Clark, made its way down the Columbia River
in the fall of 1805, the explorers were more impressed with the extent
of native salmon fishing than with the fish themselves. After crossing
the Rocky Mountains, the near-starving expedition traded with the Nez
Percé for dried salmon. Unaccustomed to this rich, oily diet the Corps
of Discovery immediately got bogged down as its members suffered
from dysentery. Though they traveled down the Columbia during the
fall salmon runs when the river was full of fish, the explorers repeatedly
traded with the locals for dogs to eat. The natives must have thought
these strange visitors were crazy.

The native population of the Pacific Northwest was less than a hundred thousand at the time of first contact with Europeans. About ten thousand people are thought to have lived in western Washington. Salmon outnumbered people by at least a thousand to one.

Even though salmon accounted for most of the native diet, the total annual catch was no more than a few million fish across the region before European contact. Though it is presumptuous to interpret, reconstruct, or evaluate Native American fisheries management after so much has changed in the region over the past 150 years, we can say with some certainty that the role of salmon in Native American cultures certainly differed from that in the region today. Several million people now live in and around Seattle alone, and the population of Washington State is projected to reach 6 million in the near future. Today there are more people than wild salmon in the state.

Native American salmon management was oriented around river systems and their watersheds because the geography of river systems organized settlement patterns. Salmon were abundant at river mouths and along major rivers. So villages tended to be located at major river outlets or good fishing sites, such as those on Puget Sound and along the Columbia River. Native Americans limited harvests and allowed large numbers of spawners to return to the rivers. Most native cultures in the region prohibited disturbance of spawning salmon, often levying stiff penalties for transgressions.

Above almost all else, native peoples valued access to their salmon fishery. Salmon were essential. In negotiating the loss of their ancestral lands, the Puget Sound and Columbia River tribes insisted on preserving their fishing rights and access to fishing grounds. Presumably, they believed that such access implied continuation of fish runs capable of being fished. Tribal negotiators were suspicious of both treaties and the government agents coercing them to sign them, but they could not have imagined that wild salmon could be nearly exterminated in little more than a century. The legacy of the Indian treaties means that the modern political landscape surrounding salmon-recovery efforts includes not only the directives of the Endangered Species Act but also interpretation of federal treaty obligations.

Native Americans came to the Pacific Northwest in several waves of immigration, the first over 10,000 years ago when glaciers were retreating from North America. Diets of the earliest arrivals were quite varied in coastal areas, and dependence on salmon increased inland from the coast. Intensive salmon fishing on the Columbia River dates from at least 8,000 to 9,000 years ago. Salmon bones recovered from archaeological sites in southern British Columbia indicate that salmon were an important part of the native diet 6,000 years ago. Great quantities of salmon bones in the upper layers at these sites indicate that native cultures relied on salmon even more once sea level stabilized about 5,000 years ago. From that time until the late nineteenth century, salmon habitat reached its maximum extent and quality in the Pacific Northwest.

Evolving native cultures developed greater dependence on salmon as the abundance of salmon habitat, and presumably salmon, increased in the postglacial world. Many salmon runs not only survived being the targets of native fisheries for thousands of years, but actually thrived under human predation. Fishing nets dating from 3,000 to 500 years ago attest to the antiquity of salmon fishing in western Washington and British Columbia. Native American cultures also evolved and adapted to dependence on salmon runs. The various species of Pacific salmon evolved along with the landscape as well as the Native American cultures and communities of the region.

Sediment cores extracted from lakes on Alaska's Kodiak Island reveal evidence of broad swings in the prehistoric abundance of salmon. After spawning, sockeye that return to the lakes die and their rotting carcasses leak nitrogen that becomes incorporated into fine sediments on the lakebed. The relative abundance of the isotope nitrogen 15 (^{15}N) in the lake sediments indicates the relative proportion of nitrogen derived from marine sources. So if you pull a sediment core up from a lakebed, the ^{15}N concentration in each layer provides an estimate of the amount of nitrogen imported from marine waters by salmon or other anadromous fish. Date the layers in the core and you have a paleo-salmon meter.

The Kodiak Island lake cores record variations in salmon abundance over the past 2,000 years. Large swings in salmon abundance coincide

An engraving made in 1778 shows a Nootka woman of Nootka Sound, British Columbia, with woven hat and a cedar-bark cape.

with both archaeological records of cultural change and climate variability. Low Alaskan salmon abundance prior to A.D. 800 coincides with warm ocean conditions in the northeastern Pacific. As the northern Pacific subsequently cooled, Alaskan salmon abundance increased. At this time, the human population in the region also increased and shifted to greater reliance on salmon fishing. Sustained high abundance of Alaskan salmon after A.D. 1200 corresponds to cool periods of glacial advances in southeast Alaska. Over the past several thousand years, large-scale climate variability influenced salmon abundance and thereby affected the development of native cultures.

Shorter-term changes also affect salmon abundance. In particular, a twenty to thirty-year cycle in ocean conditions in the North Pacific, called the Pacific Decadal Oscillation, corresponds to dramatic shifts in salmon production in Alaska and the Pacific Northwest. Curiously, changes in Alaskan salmon abundance are out of phase with salmon abundance in the Pacific Northwest and California.

When Alaskan salmon production is high, the Pacific Northwest and California salmon runs are low, and vice versa. This pattern is caused by the effect of oscillations in sea surface temperature in the middle of the ocean on coastal upwelling and marine productivity. During cold phases in the Pacific Northwest, enhanced upwelling of cold, nutrient-rich bottom waters along the coast produces lots of krill and sustains an arctic fauna with few predatory competitors for salmon. During warm phases, in contrast, krill production is lower and coastal waters are full of predatory tropical species that compete with salmon for a smaller supply of food. Natural salmon production, therefore, has a boom-and-bust cycle tied to long-term shifts in ocean temperature and the productivity of coastal waters.

Over the past century the slowly varying pattern of ocean temperature that affects marine productivity, and salmon abundance, has reversed several times. In the Pacific Northwest, conditions were bad for salmon from 1925 to 1947, good from 1947 to 1977, and then bad again from 1977 to about 2000. Right on schedule, marine conditions have improved dramatically for Pacific Northwest salmon over the last several years, indicating that a new shift to better marine conditions has probably started. With this new switch we should expect cooler ocean temperatures and high marine productivity in the near future.

How could one reliably catch a consistent number of salmon from a population of fish characterized by large natural swings in numbers? Native American systems of salmon management imposed cultural restrictions on harvests that ensured they took less than half the available fish. In contrast, modern Euro-American salmon managers harvested as much as 90 percent of the runs, a practice that has not proved sustainable. Large natural variations in salmon abundance mean that the only safe long-term strategy for sustaining salmon fisheries is to limit harvests to a fraction of the overall run size—the Native American model of salmon management.

The development and evolution of salmon fishing by the Lummi tribe in northern Puget Sound provides interesting contrasts and parallels to the role of salmon in aboriginal cultures on the Atlantic coast, where salmon were not as central to the culture and livelihood of most native peoples. Around northern Puget Sound salmon formed the basis of a way of life. Daniel Boxberger, an anthropologist who has worked with the Lummis, observed that whenever Lummis get together, conversation inevitably drifts toward salmon fishing. Salmon were and are the primary foodstuff and occupy a central position in Lummi spiritual and social life.

Before 1855, when the Lummis and the U.S. government signed the Treaty of Point Elliott, which created the Lummi Reservation, salmon fishing provided subsistence for a dispersed network of small, local communities. The Lummis traveled seasonally between sites for salmon fishing, shellfishing, and plant gathering. Early fishing practices benefited entire villages and clans rather than individuals, and no individual had control of the fishery or the means to exploit it. Villages consisted of independent houses, each composed of individual families united by kinship to other houses. Salmon fishing was the most important means of subsistence for the pre-contact Lummis, with estimates of per capita salmon consumption ranging from one to almost two pounds per day. At the time of the treaty the Lummis had twenty-six houses dispersed around North Puget Sound, with a total population estimated at 700 to 800 individuals. Hence, the pre-contact fishery in the Lummis' territory caught a quarter of a million to a half a million pounds of salmon. If each fish weighed about ten to twenty pounds, this would account for an annual harvest of somewhere between 10,000 and 50,000 salmon.

A typical house consisted of an owner and his sons, brothers, and male cousins and their families. Allegiance fell to the house in which one resided, but there was substantial intermarriage among houses throughout northern Puget Sound. This resulted in a complex social web among settlements in the region. The primary bond between the aboriginal inhabitants of the Puget Sound was a shared language and culture. Political units that we know as tribes developed only after the treaties were signed and native communities were relocated to reservations.

The aboriginal economy included free access to fishing locations within a family's kinship network, which for most Lummi families extended throughout much of northern Puget Sound. Although access to fishing was generally open to all, some locations were held in trust by certain individuals for a larger kin group. In particular, fish weirs, traps that spanned rivers, and reef nets in coastal areas could be used only with the permission of the owner, unless one had contributed labor and materials toward their construction. Fishing with methods that did not require extensive labor was unrestricted. Anyone could troll for salmon in a river or on Puget Sound.

Though ownership of reef net sites was an inherited right, the owner was obliged to select crew members from his extended family and housemates. Fish were divided among the crew, and the owner of the site traditionally ensured that the needs of the crew were met before his own. After the crew had enough fish, the owner took the rest. According to Boxberger, the literature does not mention what would happen if there had not been enough fish to satisfy the crew. Salmon shortages apparently did not occur.

Many in the region believed that the salmon lived in five great houses deep in the sea, one for each kind of salmon. Once a year the salmon would journey from their ocean abode to pay their respects to their terrestrial brothers, providing a gift of rich food. Shown the proper respect they would return to the rivers year after year. Like any guest, they would not come back if treated poorly.

Throughout the Pacific Northwest and as far south as the Sierra Nevada in California, the return of salmon was an important annual event marked by a ceremony in which the first fish caught was honored as the First Salmon and either was shared among community members or was ritually eaten by a shaman. In most ceremonies, the bones of the First Salmon were carefully returned to the water. Ideally, the spirit of this ambassador salmon would tell other salmon how respectfully it had been treated and encourage others to allow themselves to be captured.

Restraint was a fundamental characteristic of Native American salmon fishing. Cultural constraints on salmon fishing practices allowed for adequate protection of spawners. Interviewed for a legal case in 1942, Sextas Ward, a ninety-year-old Quileute Indian who was a child when the treaties were signed, recounted: "When the Indians had obtained enough fish they would remove the weirs [salmon traps spanning a river] from the river in order that the fish they did not need could go upstream and lay their eggs so that there would be a supply of fish for future years " (quoted in Swindell 1942, 222).

In another account of pre-treaty fishing practices from the same series of interviews, a seventy-nine-year-old Umatilla Indian named James Kash Kash, who was born about 1863, testified: "It was customary for the Indians not to catch the salmon in the tributaries until after they had spawned for the reason that they knew there would be no salmon in the future if they did not permit the females to lay their eggs to be hatched and available in future years" (Swindell 1942, 305).

Throughout the region ritualized limitations on the duration, and therefore the intensity, of fishing institutionalized safeguards against overexploiting salmon. For example, on the Northern California coast, catching salmon for general consumption was forbidden at the start of the spring salmon runs. The salmon season opened for general fishing only after a ceremonial period following ritual preparation and eating of the First Salmon. The waiting period lasted from several days to weeks.

Along the Klamath River fish weirs were built in the ten days after the first salmon passed and then were dismantled after ten days of fishing. In addition, weirs were opened each night to allow salmon to pass upstream until fishing resumed the next day. Some weirs even had open gaps to allow passage of some salmon at all times.

The anthropologists Sean Swezey and Robert Heizer described salmon fishing practices on the Northern California Coast as a cultural adaptation that provided for sustainable fishing: "The 'restraining effect' extended by ritual restrictions concerning salmon fishing appears to have been a widespread phenomenon—and there is no evidence that native populations ever seriously overfished the salmon runs" (Swezey and Heizer 1993, 323–324).

Swezey and Heizer concluded that in pre-contact times California salmon were

... a seasonally abundant and renewable commodity which required intelligent and competent organization and control of fishing practices to ensure efficient harvest. . . . [R]itual specialists directed and controlled fishing and dam building activities, regulated the opening of the salmon fishing season, and managed the use of the spawning runs. . . . The anadromous fish resource was perhaps the most intensely managed and ecologically manipulated food resource among these aboriginal societies. (1993, 327)

In many areas the Native American fishery provided a substantial return for intense effort during the salmon run. In some areas, the catch provided food to last the whole year.

Salmon were preserved by smoking or by drying to make jerky, which could keep for several years. The native population along the Columbia River dried enough salmon to support trade with people from the interior and coastal areas. Radiocarbon dates from piles of salmon bones excavated along the Columbia indicate substantial Native American salmon fishing since shortly after both Europe and North America thawed out from beneath glacial ice.

The native population along the Columbia River at the time of Lewis and Clark's voyage of discovery has been estimated as 50,000, with an annual catch of about 20 million to 40 million pounds of salmon. This catch conservatively translates into 1 million to 2 million fish. The Northwest Power Planning Council (an entity that was created by congress under the Northwest Power Act of 1980 with a mandate to implement an interstate program to protect and enhance Columbia River salmon runs) estimates that the pre-contact annual runs on the Columbia varied from 11 million to 16 million salmon. Native fishing, therefore, appears to have accounted for somewhere between 5 and 20 percent of the runs. Although just a modest proportion of the run size, this still made for a substantial fishery. Lewis and Clark reported passing more than a hundred native fishing stations on their way down the Columbia River. Even with such intensive operations, the native population sustained a society based on the fishery as old as the agricultural societies along the Tigris and Euphrates rivers in the Middle East.

Post-contact, new and exotic diseases decimated native peoples, especially as the frequency of contact with Europeans increased after 1820. By the 1850s, the growing European population surpassed and began to displace the shrinking native populations in Oregon and along the Columbia River. By 1851, the Native American population along the Columbia River fell to just one sixth the size of the pre-contact population. Furthermore, the European expansion began to separate the Native Americans from their fisheries. George Catlin, an artist who visited the region at this time, commented on the destitution of the native population along the Columbia River when deprived of access to salmon:

> The fresh fish for current food and the dried fish for their winter consumption, which had been from time immemorial a good and certain living for the surrounding tribes—is now being "turned into money," whilst the ancient and real owners of it may be said to be starving to death; dying in sight of what they have lost, and in a country where there is actually nothing else to eat. (1959, 144)

Already ravaged by new diseases, the native population was at a serious disadvantage in dealing with the new arrivals. Plans for a transcontinental railroad had been brewing in Washington, D.C., since Asa Whitney introduced a resolution in Congress in 1845 endorsing building a railroad to the Pacific. Competition was intense between southern and northern interests wrestling over the proposed routes for the railroad. Secretary of War Jefferson Davis pushed for a southern route. Isaac Stevens, who in 1853 led the surveying party charged with exploring possible routes between Michigan and Puget Sound, was the champion of the northern route. Later that year Stevens was appointed governor of the newly created Washington Territory and was also appointed superintendent of Indian affairs for the new territory. The U.S. government officially recognized aboriginal land ownership, and so in order to build the railroad, and open the territory for the settlers it would bring, the government needed to acquire title to the new territory. Stevens's primary mission was to eliminate aboriginal land ownership as quickly as possible.

In 1854 and 1855 Stevens traveled around Washington Territory negotiating a series of treaties with native groups. (In the end, ironically, given Steven's haste to grab land, the Civil War delayed plans for the railroad.) Although the twenty-six tribal reservations in the State of Washington together cover less than 8 percent of the state, the United States Senate did not ratify many of the treaties until 1859, four years after they were signed, because of concerns that Stevens had been too generous with the Indians. Today the federal government retains ownership of 29 percent of the state, and timber companies own many times the area of all the reservations combined. What did the Indians get in return for the 92 percent of the Washington Territory that they relinquished to the U.S. government in the treaties? A few cents per acre, the promise of government protection, and the right to keep fishing for salmon.

Some question how honorably Governor Stevens conducted treaty negotiations. At Stevens's insistence, the treaties were negotiated in Chinook, a trading jargon with at most a few hundred words in common use drawn from a mix of Indian languages, English, and French. Owen Bush, a member of the governor's staff, later recalled, "I could talk the Indian languages, but Stevens did not seem to want anyone to interpret in their own tongue, and had that done in Chinook. Of course it was utterly impossible to explain the treaties to them in Chinook" (Meeker 1905, 208). Bush and others also wondered how well Stevens's translator, Colonel Benjamin F. Shaw, knew the Indian languages.

But Stevens knew what he wanted. He'd had the treaties drawn up in advance. First and foremost, he wanted to clear the title to the land and concentrate the Indians in one or two relocation areas to open the territory for settlers. Stevens warned the Indians that the government could not protect them from the deluge of settlers coming over the horizon unless they relocated to reservations. Though many Indian leaders were dissatisfied, they eventually signed the treaties. Stevens told them that if they refused to sign, legions of settlers would overrun them anyway. In describing the negotiation of the Stevens treaties, William Compton Brown, a retired superior court judge from eastern Washington who interviewed Yakima tribal members in the early twentieth century, wrote that "haste, high pressure, and no little chicanery on the part of the whites was predominant throughout the meetings from start to finish" (Brown 1961, 64).

Though Stevens was authorized to relieve the Indians of their lands, he realized that there could be no agreement without assuring their continued right to fish in their accustomed places. Once reconciled to forfeiting most of their land, the primary interest of the tribes lay in securing and protecting their right and ability to catch salmon. Stevens knew that the Indians were far more willing to move and accommodate new neighbors than they were to stop fishing for salmon. The treaties uniformly included language preserving fishing rights: "The right of taking fish, at all usual and accustomed grounds and stations, is further secured to said Indians in common with all citizens of the Territory" (Minot, 1855, 1133).

No matter how well or clearly the various treaties were translated into the Indian languages, there remained a simple underlying inequity. The Indians knew that there were plenty of salmon to share and had no reason to believe there might not be in the future. There had always been plenty in the past. The government Stevens represented knew of the depletion of salmon in England and along the Atlantic seaboard. Stevens not only orchestrated the negotiations but he also knew the game they were playing.

Today this may seem like irrelevant ancient history, but the treaties still provide the legal basis for federal sovereignty over the region, as well as tribal fishing rights. I first became aware of the treaty language when Hiram Li, a biologist from Oregon State University, asked the governor's Salmon Recovery Office to provide us with copies of the treaties at the first meeting of Washington's Independent Science Panel that I was appointed to in 1999, and which was charged with reviewing the state's salmon recovery program. Taken by surprise, the governor's office arranged for a lawyer from the state attorney general's office to come to our next meeting and explain the treaties to us. What we heard convinced me that the treaties implied that government had a responsibility to preserve salmon runs through habitat protection and land use restrictions if necessary. In further discussions we quickly discovered that this was not the question that the state wanted us to be asking.

The treaties did specify that the Indians were to share the right to fish for salmon with the settlers, but did not guarantee the tribes an equal footing for competing with commercial salmon fishers. After the introduction of cannery technology made salmon exports from the region a viable enterprise, most of the fish processed by Puget Sound canneries initially came from Indian fishing. But as commercial fishing became increasingly profitable, and competitive, natives were rapidly squeezed out of business by lack of access to capital. Banks would not lend money to finance acquisition of fishing boats to reservation Indians because they had no collateral. They could not use their reservation lands as leverage because the federal government held these lands in trust. The right of access to ancestral salmon fishing sites did not guarantee the ability to participate in commercial fishing.

Lacking the capital needed to apply their skills to the developing commercial salmon fishery, Native Americans were outpaced by technological change in their ancestral industry. In short order, large numbers of white-owned salmon traps patterned after those used for Atlantic salmon on the East Coast began to displace native fishermen. These traps were ingenious devices consisting of a submerged fence of twine or galvanized wire mesh connected to piles driven into the bed of Puget Sound or a river. Fish were directed into the maze through openings in the structure that guided them into progressively smaller pens from which they could be scooped with dip nets. Such traps were extremely efficient, cost-effective, and profitable. Those in place in 1900 yielded lucrative profits that could more than double an investment in a single season.

Naturally the traps multiplied. In the first decade after their introduction, the number of fish traps on Puget Sound grew tenfold. Many cannery-owned traps were located on, near, or directly in front of traditional native fishing sites. At Point Roberts, for example, Indian fishing accounted for less than 5 percent of the salmon harvest by 1895. A perennial source of contention, fish traps were outlawed in Oregon in the 1920s and in Washington in the 1930s.

As fish traps declined, power boats came to dominate commercial salmon fishing on Puget Sound. New vessels towed nets that could be drawn together like the drawstrings on a purse, trapping everything inside. By 1935 the State of Washington licensed almost a thousand power fishing vessels annually. Native Americans had become all but excluded from commercial salmon fishing.

Just as important as the lack of access to capital and the rapid technological change in the fishery was the political situation, as the federal Bureau of Indian Affairs aggressively sought to reduce native salmon fishing. Bureau policy stressed agricultural development and discouraged fishing. The State of Washington also worked to curtail Indian fishing. Given the central place that salmon fishing occupied in native cultures of the region, it is not surprising that efforts to curtail Indian fishing proved controversial. Expansion of the commercial fishing industry and then competition for a declining salmon harvest led to conflict over the meaning of the Indians' exclusive right to fish within tribal reservations, and especially their right to fish off-reservation at their usual and accustomed places. For a century much of the State of Washington's effort at restricting salmon fishing was directed at tribal fishermen, even as changes in technology and fishing gear progressively reduced the proportion of the salmon harvest taken by Native Americans.

Beginning in 1894 the Lummis petitioned the federal government to protect their access to fishing grounds from encroachment and displacement by white fishermen. In 1897, C. H. Hanford, a federal judge with ties to the local fishing industry, ruled that under the Treaty of Point Elliott native fishermen retained equal rights but did not have special privileges protecting their access to particular locations over the interests of other citizens. The ruling encouraged the state to restrict Indian fishing. To this day, the no-special-privileges argument pops up on talk radio and in state politics.

Some abuses of Indian treaty rights were flagrant enough to spur federal action. In the 1890s two brothers, Lineas and Audubon Winans, operated a state-licensed fishing operation on homesteaded land near Celilo Falls. The Winans brothers forcibly prevented Yakama Indians from crossing their land to reach their traditional fishing grounds. The Winanses must have been shocked when the local U.S. attorney charged them with violating the treaty signed by Yakama tribal leaders and Governor Stevens in 1855. The Indians lost in the lower courts, which ruled that the treaty only guaranteed Indians rights equal to

Columbia River area Indians fish with spears at Celilo Falls, Oregon, ca. 1910.

those of white citizens, and therefore that their treaty-protected right to access fishing grounds could be superseded by state licenses and federal homestead grants. An appeal of the lower-court ruling landed before the U.S. Supreme Court in 1905.

In a scathing opinion, the Supreme Court backed the Indians.

> [I]t was decided [in the lower court] that the Indians acquired no rights but what any inhabitant of the territory or state would have. Indeed, acquired no rights but such as they would have without the treaty. This is certainly an impotent outcome to negotiations and a convention which seemed to promise more, and give the word of the Nation for more. . . . [T]he treaty was not a grant of rights to the Indians, but a grant of rights from them. . . . No other conclusion would give effect to the treaty. (*United States v. Winans*, 1905, 380–382)

The Supreme Court ruling further held that the transition from territory to statehood did not nullify Indian fishing rights. Nonetheless, the ensuing legal battles over the scope and nature of Indian fishing rights focused on the tension between federal treaties and state laws.

From the beginning of this legal war, federal courts upheld and protected Indian fishing rights, while state courts curtailed or limited those rights. State court defeats for Indian fishing rights were generally overturned when appealed to federal court. Still, the fight dragged on for a century.

One of the first regulations enforced by the State of Washington's Department of Fisheries, created in 1890 to regulate the salmon and other fisheries, was to ban the use of nets in streams. By then this had become the primary method of Indian fishing because traps built and operated by canneries had displaced Indian fishing from coastal areas up into rivers and streams. The Department of Fisheries took this step because the Indians' river fishing, now in plain sight of the general public, undermined state fisheries managers efforts to protect the few fish that made it past the commercial traps. This arrangement guaranteed conflict over Indian fishing by placing Indians last in line, and squarely in the public eye, in the fight over a declining resource. Public anger over Indian fishing grew, and the state began to enforce fishing regulations on the reservations even though the treaties protected on-reservation fishing from state regulation.

As commercial fishing began to deplete salmon stocks, enforcement of regulations occurred mostly in the rivers, rather than in open water. Because Indians lacked access to capital to buy commercial fishing boats, Native American fishing remained relegated largely to rivers. Regulatory practices compounded the disadvantage the Indian fishermen faced. Fishing seasons were timed to coincide with open-water runs and to end when salmon entered the rivers, leaving the Indians no time or place to fish. Refusing to allow even a subsistence Indian fishery, the state announced it would arrest any Indians fishing in defiance of state law—whether on or off their reservations. The manner in which the state regulated salmon fishing helped turn public opinion against Indian fishing.

Publicity surrounding tribal defiance of state fishing regulations fueled longstanding perceptions of Indian fishing as contributing to declining salmon runs. Yet in 1935, the first year that the state kept records, Indians caught less than 2 percent of the total salmon catch. The powerboat fleet hauled in 90 percent. According to state records, the entire Indian catch for Puget Sound from 1935 to 1950 accounted for less salmon than taken by the commercial fishing fleet in one typical year during the same fifteen-year period. Nonetheless, the state continued to blame deterioration of salmon stocks on Indian fishing. For example, in 1959 the Washington State Department of Fisheries blamed Indians for undermining salmon-recovery attempts, stating, "The effort on most Indian fishing streams has increased tremendously over the past several years and the results of many a protective regulation on commercial and sport fishing has served merely to fill an Indian net" (Washington State Department of Fisheries 1959, 221).

Although the state directed substantial effort toward censuring Indians for their perceived role in depressing salmon runs, it did not limit entry into the commercial fishery of Puget Sound, and thus the fishing fleet continued to grow. In addition, the state continually maximized the allowable harvest by the state-licensed commercial fishermen and did not allow for any river-based fishing. Consequently, the Indians took salmon after the full catch allowed by fisheries managers had already been harvested. Instead of accounting for Indian harvest in their planning, which would have required some commensurate lowering of the limit for commercial fishing, the state cultivated public support to eliminate Indian fishing. By 1961, the state acknowledged the perilous state of its salmon but continued to emphasize control of the Indian fishery.

> It does no good to talk of how many salmon there used to be, or that Indian treaties supposedly guaranteed Indians the right to take fish in perpetuity, or argue about who or what has reduced those salmon runs to their present state. The important thing is that our salmon runs are now dangerously low and in many cases are even to the point of facing extinction in the very near future. (Washington State Department of Fisheries 1961, 178)

Despite such prescient rhetoric, the Washington State Department of Fisheries continued to grant increasing numbers of commercial fishing licenses. The number of gill-net licenses issued by the state increased fourfold between the early 1950s and the 1970s. The sport fishery in Puget Sound also mushroomed during this time. By the early 1960s more than half a million salmon were being taken annually by the recreational fishery, which by then had surpassed the Indian fishery.

In 1970, the federal government on behalf of western Washington tribes that it had signed treaties with sued the State of Washington, alleging that commercial salmon fishing had come to exclude treaty-protected fishing by Native Americans. The landmark case, and the subsequent opinions that it produced, bear the name of Judge George Boldt, a no-nonsense federal district court judge who ruled that the treaty language "in common with" meant that the treaty tribes were to share equally with non–Native American citizens of Washington State the opportunity to take fish at their usual and accustomed places. Judge Boldt's 1974 ruling, which was upheld by the U.S. Supreme Court in 1979, held that the treaty tribes were entitled to half of the salmon harvested in their traditional fishing areas.

The initial Boldt decision was a bombshell. The idea that the salmon harvest was to be shared equally between the treaty tribes and the other citizens of the state of Washington was unexpected. The fishing industry was stunned. Indian fishermen had previously taken a small fraction of the catch, and now they were entitled to half of all the salmon harvested in the state. Violence erupted as non-Indian fishermen fumed that the Boldt decision would destroy their livelihood. Few wanted to admit that the collapse of commercial salmon fishing was just a matter of time anyway as long as salmon runs continued declining as a result of overfishing, habitat degradation, and the blocking of rivers and streams. In the end, the tribes finally secured their treaty rights just as the commercial fishery for wild salmon virtually expired.

Because Indian fishing was such a small component of the overall fishery at the time of the Boldt decision, many in the general public believe the decision as *giving* half of the fish to the tribes, and the media often portray the issue the same way. But the court concluded that through the treaties the tribes originally *reserved* the right to half the salmon and *granted* the citizens of the United States the privilege of taking the other half. In 1905, the U.S. Supreme Court in ruling against the Winans brothers had confirmed that federal treaties supersede state law, and finally almost a century later the issue of Indian fishing rights appeared to be settled.

Or was it? The tribes, state, and federal government returned to court dozens of times in the four years after the initial Boldt ruling. During the appeals and challenges, federal courts ordered state agencies to implement the Boldt decision, but state politicians instructed state agencies to manage salmon against federal directives. After four years, following what one appellate judge characterized as one of "the most concerted official and private efforts to frustrate a decree of a federal court witnessed in this century," Judge Boldt assumed jurisdiction over the salmon fishery and the Coast Guard was called in to enforce federal fishing regulations (*Puget Sound Gillnetters v. United States District Court*, 1978, 1126). The next year, the U.S. Supreme Court upheld the Boldt decision and the state was forced to recognize the treaty tribes as comanagers of the salmon fishery. After the Boldt ruling, Congressman Lloyd Meeds and Senator Slade Gorton of Washington repeatedly introduced bills to reduce or eliminate tribal treaty rights.

When he adjudicated the case brought by the federal government over tribal fishing rights, Judge Boldt separated the issue of protecting the habitat necessary to produce salmon from whether the treaties guaranteed the tribes a particular share of the harvest. Having ruled in favor of a tribal right to half the salmon harvest, Judge Boldt died before settling the habitat question. The issue ended up before the Honorable William Orrick, another federal district court judge. In his 1980 ruling Judge Orrick stated, "The most fundamental prerequisite to exercising the right to take fish is the existence of fish to be taken." He ruled that the treaties provided the tribes with the guarantee that state and private interests must refrain from degrading salmon habitat to an extent that would deprive the tribes of their "moderate living needs." (*United States v. State of Washington*, 1980, 203)

Fearing wholesale erosion of their ability to permit land uses that harmed salmon, the state appealed. An eleven-judge panel of the Ninth Circuit Court of Appeals in San Francisco vacated Orrick's ruling, deciding that it was improper for the court to rule on the habitat issue without an actual on-the-ground controversy over some project or particular action. Deciding the appeal on a procedural technicality, the panel did not address the legal basis of Orrick's conclusion. The issue remains unresolved.

Non-Indians sometimes complain about Indians' being granted special rights. But unlike emigrant communities that came to America voluntarily, Indian tribes are recognized by the federal government as sovereign nations with whom the government negotiated international treaties. Those treaties are still in effect, and they provide the legal foundation for incorporating much of the western states into the United States. Furthermore, as Judge Boldt and others have noted, it is the Indians who granted rights to the federal government, not the other way around.

Some state politicians and citizens may feel that Indian treaties that predate the existence of the state shouldn't carry much weight. But it is the federal government's obligation to honor the treaties. The State of Washington can no more unilaterally abrogate the Stevens treaties than it can revoke the Anti-Ballistic Missile Treaty or the Geneva Conventions on the conduct of warfare. No matter how much situations or conditions change, obligations under these treaties do not change with the political winds.

Tribal treaty rights are an emotional, volatile issue that will likely be fought over for years to come. How these issues play out will have a central role in the future of salmon. Native American salmon management in the Pacific Northwest was practiced historically on a river-by-river basis and was characterized by institutionalized restraint to ensure that relatively high numbers of fish would make it back up a river to spawn. This combination helped sustain native fishing for thousands of years. By the time Governor Stevens secured his treaties, this was not the case in the homeland of my ancestors, where salmon were already in trouble and a salmon crisis was raging. The fall of Pacific Northwest salmon had clear precedents half a world away.

OLD WORLD SALMON

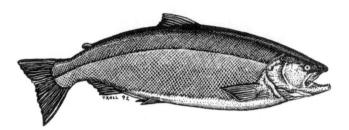

The River Salmon surpasseth all the fishes of the sea.

Pliny the Elder (A.D. 23–79)

SALMON CAPTURED THE HUMAN IMAGINATION LONG BEFORE they acquired the name we know them by today. Europeans living on the edge of the ice sheet at the peak of glaciation painted images of salmon on cave walls and carved images of salmon into reindeer horns. Who really knows whether such depictions were rooted in mystical respect, aesthetic appreciation, or the importance of salmon as food (or all three)? Whatever the reasons behind their art, early Europeans relied on salmon long before Europe emerged from under glacial ice.

The Latin name for salmon, *Salmo*, has grown by just one letter in two thousand years, but our understanding of these fish has evolved dramatically. Although little was known in classical times about the natural history of salmon, Caesar's legions marveled at the hordes of big fish leaping in the rivers of Gaul (better known today as France). Yet the Romans did not see salmon solely as entertainment. As the empire spread north, salmon became a favorite item in Roman markets.

Early curiosity about salmon focused on how they could hurl themselves out of the water to climb waterfalls. In fact, the generic name *Salmo* is derived from the Latin word *salir*, meaning "to leap." One long-held popular misconception was that by taking its tail in its mouth a salmon could rotate up and over seemingly impassable falls. Confusion over the basic ecology, habits, and life history of salmon continued until relatively recently. The eighteenth-century taxonomist Carl Linnaeus gave the Atlantic salmon the scientific name *Salmo salar*, which, depending upon whose interpretation you believe, means either "salmon the leaper" or "leaper from the sea" (*salarius* means "of the sea"). It took another century to sort out the life history and habits of the Atlantic salmon, and until just a decade ago to finalize the salmon's evolutionary family tree.

Rivers across northern Europe were once full of salmon. Until the 1700s thousands of Atlantic salmon were taken each day during spawning runs on French rivers. But by the mid-1800s continental stocks were so depleted that the French government instituted an aggressive (though ultimately futile) restoration program. Huge 50-pound salmon ran up the Rhine River until after the unification of Germany in 1871 and reached well into the continental interior until the late 1800s. Up to a quarter of a million salmon were caught from the Rhine each year before twentieth-century industrial pollution wiped out the Rhine salmon. By 1960, salmon were extinct in Germany, Belgium, the Netherlands, and Switzerland. Today salmon have been all but exterminated from the rivers of Spain, France, Portugal, Denmark, Finland, and the Baltic states. Just four European countries—Norway, Ireland, Iceland, and Scotland—still have comparatively healthy populations of salmon. Only remnants remain of continental Europe's once mesmerizing salmon runs.

The decline of European salmon foreshadowed the demise of salmon in New England and the Pacific Northwest. Writing in the eighth century, Saint Bede (known as the Venerable Bede) took note of a plentiful supply of salmon and eels in Britain. The Domesday Book, compiled for William the Conquerer in 1086, twenty years after the Norman Conquest, recorded among England's notable resources an abundance of salmon. For example, it was recorded that the lord of the manor at Gloucester collected a thousand salmon a year in tribute from fishermen. In subsequent centuries, a long series of royal proclamations and regulations indicate sustained concern over conservation of salmon fisheries. These measures show that what was needed to keep the rivers full of salmon was understood reasonably well, even though knowledge of the natural history of salmon remained sketchy. Quite simply, it was perfectly obvious that to keep a river full of salmon, enough adult salmon had to reach their spawning grounds, and enough juvenile salmon had to reach the sea.

One legacy of early confusion over the natural history of salmon is the variety of names for the salmon's different life phases. "Fry" and "parr" both refer to juvenile salmon smaller than the larger juveniles, called smolts, that are ready to migrate to the sea. There are also a number of different terms for adult Atlantic salmon. Most Atlantic salmon spend two to three years feeding and growing at sea. But some return after a single year. These small yearling salmon are known as grilse. Until the 1860s many fishermen erroneously considered grilse a separate species.

The few adult salmon that survive after spawning are known as kelts. Exhausted but not finished, they drift downstream to slow-water pools, where they rest until high water flows propel them back to the sea. The proportion of kelts in a run that survive to spawn again varies from about 5 percent to as much as a quarter of the run. Less than 1 percent survive to spawn a third time and less than one in a thousand make it to spawn a fourth time. So in the vernacular of British fishermen a parr or fry becomes a smolt that heads to the sea to grow into a grilse after its first year. Only after a fish has spent several years growing at sea do the English call it a salmon.

Salmon fishing may be the oldest ongoing regulated profession in the English-speaking world. Regulation of the salmon rivers and fisheries of Scotland began before the Norman invasion. King Malcolm II, an early

Scottish king, in 1030 issued the earliest known restriction on salmon fishing, an edict establishing a closed season for taking old salmon between Assumption Day, at the end of August, and Martinmas, in mid-November. The Magna Carta, presented by the English barons to King John and signed by him in 1215, provided for the dismantling of the king's fish weirs in order to protect salmon for use as a public good throughout all England. Seventy years later the Statute of Westminster formalized the lasting basis for English fishing law by codifying regulated fishing seasons. The law, signed by Edward I in 1285, specified severe penalties for salmon poaching—fishing out of season or without permission in private waters that escalated for repeat offenses. First offenders risked losing their fishing gear, second offenders spent three months in jail, and habitual third offenders faced a year in the dungeon.

Fishing rights were originally reserved for the Crown, but in the wake of the Magna Carta England's tidal waters were opened to public fishing. On non-navigable waters fishing rights went with ownership of land along streambanks, and therefore were retained by the landowning nobility. Landowners were free to rent access to their fishery, subject to regulations prescribing fishing seasons and the permitted types of gear.

Regulating the effect of dams and other in-channel barriers to fish migration also has a long history. A statute dating from the reign of Richard I (the Lionhearted) in the twelfth century declared that English rivers be kept free of obstructions so that a well-fed three-year-old pig could stand sideways in the stream without touching either side. This mandate to leave a hole in anything built across a river was intended to allow adult salmon to reach their spawning grounds and these fish passways came to be known as the King's gap. An act of Parliament passed in 1318 during the reign of Robert I (Robert Bruce) forbade the erection of fixtures that would prevent the progress of salmon up and down Scottish rivers. The penalty for impeding salmon passage carried the penalty of a stiff fine and forty days in prison.

Statute after statute aimed at protecting the fishery forbade taking juvenile salmon, restricted fishing seasons, and imposed stiff penalties for blocking fish migration. James I of Scotland banned smolt traps and weirs that blocked fish migrations in 1424 and set the closed season for Scottish rivers as beginning on August 15. This act remained in force for over four hundred years, until it was repealed in 1828. Early monarchs apparently did not consider the possibility that their decrees might be rescinded. Scotland's James III declared that contrivances that killed juvenile salmon "be put away and destroyed forevermore," and by 1504, seventeen parliamentary acts concerning salmon had been passed in Scotland. In England, in the fourteenth and fifteenth centuries, Henry IV, Henry V, and Henry VI dictated statutes designed to curb the practice of blocking rivers either to collect salmon or for any other purposes in ways that would block migrating salmon.

An exception to the generally protective intent of the early Scottish salmon acts was an act passed in 1429 by the ninth parliament of James I authorizing Scotsmen to poach salmon in English waters. It also authorized successful poachers to export the catch overseas if the English would not buy their fish back.

Local courts or fish wardens oversaw implementation and enforcement of all these royal decrees, but compliance with early fishing laws and regulations was uneven. Ranulf Higden, a fourteenth-century monk, observed in his history chronicle, *Polychronicon*, that England was "rich in noble rivers with plenty of . . . salmon and eels. So that the churls in some places feed fish to sows." The practice of feeding young salmon to pigs was illegal, but enforcement was lax and magistrates lenient. In his *Booke of Fishing* (1590) Leonard Mascall records that laws were often violated or ignored. Mascall complained of the destructive methods used to catch juvenile fish and noted that those entrusted to guard river fisheries did not aggressively enforce regulations. He described how fishermen and fishery owners alike had little care for preserving the common wealth, favoring instead to increase their personal wealth. Mascall also noted that salmon were so plentiful in Irish streams that the duke of Ormond hunted them with dogs.

Fortunately, the salmon had powerful allies. Protecting the salmon was an important role of the monarch—the ultimate authority in the land. In 1400, King Robert III decreed that three convictions for killing salmon in forbidden times constituted a capital offense. In 1581, James

VI, adopting a system patterned after one already in place for English rivers, appointed conservators to watch over Scottish rivers. Charged with enforcing salmon conservation laws, these Crown-appointed river keepers were independent advocates for the resource who helped prevent depletion of English salmon runs despite irregular enforcement of fisheries regulations. They could jail offenders and levy fines capable of bankrupting all but the wealthy. An act passed in 1597 made landowners adjoining salmon rivers responsible for preventing salmon poaching on their lands during the closed season.

And strangely enough, England's land-tenure system also helped protect salmon. Many of the best salmon streams were located in hunting and fishing preserves on aristocratic estates. Landholders usually rented out access to their fishery, and often left much of their land more or less undisturbed. Over the centuries, large estates protected streams as the English countryside was transformed into a network of farms and ever-larger towns.

Salmon were a primary food source for people throughout the British Isles. Meat was expensive, salmon were cheap. It was easy to catch salmon when hordes of them ran up rivers and creeks to spawn. They fended for and fed themselves and then swam up shallow streams where, with a little effort, anyone could catch one. Salmon were so common as to become disdained as a meal. English and Scottish lore are replete with tales of labor agreements forbidding the feeding of salmon to servants and apprentices more than three times a week. Whether apocryphal or not these tales show that salmon were a common meal throughout the British Isles. The combination of low human population density, relatively inefficient fishing methods, and sound laws and regulations aimed at ensuring salmon conservation prevented the overexploitation of this important public resource.

Well-intentioned conservation laws were based on what little was then known of the natural history of salmon. Twenty-five years after Columbus first set foot in the New World, Hector Boethius published his *History of Scotland*, in which he remarked upon the mysterious observation that smolts would head to sea and return within twenty days, having grown manyfold but with nothing in their stomachs. Although it

Major salmon rivers of Great Britain.

didn't take long before naturalists realized that spawning salmon had actually gone to sea a number of years earlier, other basic misunderstandings of the life history of salmon remained widely held. Centuries after Brother Higden's lament, young salmon still were fed to livestock based on the common misunderstanding that parr (and fry) were a separate diminutive species.

Slowly, knowledge of salmon accumulated, although much remained speculative and widely debated. Sir Francis Bacon reported that salmon could hear and smell, and thought that they could reach ten years of age. Izaak Walton's *The Compleat Angler*, published in 1653, reported fish tagging experiments that demonstrated that young salmon went to sea and then returned to the river where they originated. In the late seventeenth century, there was substantial interest in salmon fed in part by debate over relations among different life stages. Only smolts ready to run to the sea were generally recognized to be juvenile salmon. Consequently, many fisheries laws and early conservation efforts overlooked the earliest life stages of salmon. Although scientific knowledge of salmon did not really exist to guide regulations, fishing intensity was low enough so that even loosely enforced laws were capable of protecting salmon runs.

Commercial salmon fishing was not a major export business during the Middle Ages owing to the difficulty in preserving fish during transport to market. Pickled salmon was shipped to Flanders and France as early as 1380, but the export fisheries of the British Isles only grew as new methods of preservation fed increased demand. Scottish salmon for export to England and continental Europe were preserved in barrels topped with brine, or were smoked or dusted with crushed salt. In most areas, however, there was little commercial demand for fresh salmon beyond country markets, where it sold for a penny or two per pound. Fishermen were said to complain more often about bursting their nets by trying to haul in too many fish than about lack of salmon.

Large-scale commercial trade in fresh salmon began in 1786 when a Scottish merchant, George Dempster, began shipping Scottish salmon packed in ice from the River Tay four hundred miles south by fast carriage to the markets of London. Not only did the fish arrive in excellent condition, but they could be preserved (and, more important, sold) for a week in hot weather—and for longer in cold weather. The idea caught on and the value of Scottish salmon fisheries skyrocketed as the development of railroads and steamboats accelerated the flow of fresh salmon into growing industrial cities. Iced salmon started flooding into London from the chief salmon-producing rivers of Scotland and Ireland. By

1817, more than 750,000 pounds of Scottish salmon was reaching London annually. Across England, local demand in urban centers created an economic vacuum that sucked the salmon catch toward places where salmon already were depleted, if not on the verge of extinction.

In the first half of the nineteenth century, naturalists with knowledge of salmon rivers became alarmed by the effects of overfishing and the rapid spread of stream blockages that prevented salmon from reaching their spawning grounds or returning to the sea. J. Cornish, a member of the English gentry who authored a treatise on the state of the salmon fisheries, wrote: "The salmon is one of the most valuable fish we have; yet . . . mankind seem more bent on destroying the whole race of them than that of any other animal, even those which are most obnoxious. Of this there cannot be a stronger and more conclusive proof than their present scarcity, contrasted with their former abundance" (1824, 4–5).

Although many details remained undiscovered, the basic aspects of the natural history of salmon were by then well known. Moreover, the reasons for the decline of salmon fishing in English and Scottish rivers were readily apparent.

Listing the laws passed since the time of Edward I, Cornish noted how many of the existing fishing devices, such as river-spanning nets, as well as the mill dams that now blocked rivers and streams had been illegal for centuries. He recounted how streams abounded with fish locks through which few salmon were allowed to pass. In his opinion, a key problem lay in placing enforcement authority in local county justices of the peace, characters well known for looking the other way. Cornish was appalled how fish passage blockages, "contrary to the express letter of so many acts of parliament, should ever have been erected; and still more, that they should have been continued and tolerated for so many years. . . . [T]he healing hand of time has sanctified their illegal creation" (Cornish 1824, 174).

After a while the simple existence of a dam became enough to justify its continuance as a fish blockage. If illegally constructed dams and fish passage barriers eventually became legal, then over time more and more rivers and streams would be blocked and salmon numbers would be reduced commensurately.

Cornish thought that solving the problem of declining salmon runs was not terribly complicated, that "common honesty and common sense" were all that was needed to figure out what needed to be done. He maintained that all that was necessary to fix the sad state of the salmon fisheries was to remove obstructions that blocked salmon migration (both upstream and downstream), protect salmon while spawning, and protect salmon fry and smolts until they reach the sea. Cornish further recommended appointing a sufficient number of conservators to enforce the laws on blockages and closed seasons. With adoption of these simple measures, Cornish felt that the salmon crisis could be readily solved and salmon would again fill British rivers.

Almost a decade after Cornish published his views on the state of the salmon, Alexander Fraser, a Scottish fisherman and naturalist from Inverness, published a summary of the knowledge of the natural history of salmon with the aim of encouraging better management of Scottish fisheries to provide cheap food for the poor. Describing experiments performed by both himself and acquaintances, Fraser maintained that salmon faithfully returned to their river of origin, whether guided by blind instinct as most maintained or by smell, as Fraser suspected. Experiments in different rivers conducted by numerous experimenters who marked individual juveniles by tying wire around the tail of grilse or removing a fin or part of the tail from smolts or fry, all found that salmon returned to the stream where they had been born. For Fraser the fact salmon returned to their native rivers after their migration to the sea was established beyond all doubt.

Fraser also discussed other basic aspects of the natural history of salmon at length. He noted how runs in different rivers and also salmon from different parts of the same river spawned at different times. Larger fish spawned in larger streams, whereas smaller salmon spawned in smaller streams or in slow water near streambanks. Fraser noted how it was well known that salmon spawned in shallow riffles rather than the deep pools in which they rested. He described how salmon dug their nests in the streambed, and then after spawning turned and used their tails to cover the fertilized eggs. He described how eggs develop buried 8 to 12 inches deep for three months, to emerge as wary fry intent on descending to the sea. Fraser noted that salmon would subsist on little or no food between entering the river and returning to the sea and also that salmon tend to wait at the mouth of rivers or streams to run up during high flows. He discussed their keen sense of smell and feeling, but questioned the assertion that they could hear. Fraser explained the ancient mystery of how salmon can leap 8 to 10 feet out of the water by comparing the articulated skeleton of a full-grown salmon to a steel spring. The mysterious ways of the salmon were beginning to be deciphered.

By the 1860s, the nature, source, and extent of mortality at each life-history stage were well known, as were the timing of upstream and downstream migrations and the relationship of different life-history stages to each other. There was no more mystery to the effect of temperature on the time salmon remained in fresh water before going to sea, or the time needed for eggs to develop in streambed gravel. Not only were salmon understood to return to their home stream, but some naturalists even maintained that salmon returned faithfully to spawn in the very same gravel bar of their point of origin.

Although the basic habits of the Atlantic salmon in rivers and streams were now understood, the location of its oceanic feeding grounds remained a mystery until late in the twentieth century, in part because Atlantic salmon from North America, Iceland, and parts of Europe converge on Greenland to feed and thus escaped observation. One theory to explain this behavior is that it preserves ancient patterns of oceanic migration dating back in geologic time to when the Atlantic Ocean was narrower. In this view, ancestral salmon progressively made longer and longer journeys to their feeding grounds as the Atlantic Ocean widened. Whatever the reason, Atlantic salmon travel up to 2,500 miles across the ocean to feed off krill, tiny crustaceans that give salmon their red-colored flesh, and small fish in the waters off western Greenland. Most impressive is that in spite of the energy expended in this epic journey salmon grow enormously, leaving their home streams as smolts weighing just ounces and returning several years later as adults weighing more than 10 pounds.

Recognition of the role of habitat quality on salmon abundance developed more slowly than recognition of the role of dams and overfishing, perhaps reflecting that at the dawn of the modern era, the rivers and streams of the British Isles were already far from pristine. By the seventeenth century, the English Navy's need for masts focused national attention on the wholesale destruction of English forests. John Evelyn called for protecting the remaining English forests due to the nearly complete clearing of the original forest. But much of England's oldest forest in Evelyn's day had regrown since being cleared during Roman times. Deforestation occurred so far in the distant past that geology and archaeology provide the only direct record of the original conditions in most of England's salmon rivers.

The fragmentary evidence available to us today suggests that over the centuries forest clearing transformed many British rivers. The form of the Gearagh, a reach of the River Lee flowing through a rare pocket of ancient forest near the city of Cork in southern Ireland, suggests that ancient deforestation caused dramatic changes in the character of British rivers. The river does not follow a single course but is diverted by logs and fallen trees into an intricate system of intertwined channels. This branching complex of floodplain channels resembles the forested floodplains in Washington State that still host trees large enough to deflect flow and anchor stable logjams once they fall into rivers. In contrast, modern single-channel British rivers do not exhibit such a complex form, although evidence that some, and perhaps many, once did lies preserved in their prehistoric deposits. Old European maps show many rivers such as the Rhine that once were salmon rivers as complex networks of main and side channels before their riparian forests were cleared or they were confined to a single readily navigated channel.

We are still learning how such profound changes in river form affected salmon and freshwater ecosystems. Side channels make ideal rearing habitat for juvenile salmon, as well as places for adult fish to seek refuge when violent floods roar through the main channel. The extent and timing of the historic and prehistoric transformation of Britain's forest rivers are not yet fully understood, but it is safe to say that complex, multichannel rivers provided more hospitable habitat for salmon than today's simple canal-like rivers.

Anno Regni
A N N Æ
R E G I N Æ
Magnæ Britanniæ, Franciæ, & Hiberniæ,
N O N O.

An Act for the better Preservation and Improvement
of the Fishery within the River of *Thames*, and for
Regulating and Governing the Company of Fishermen of the said River.

L O N D O N,
Printed by *John Baskett*, and the Assigns of *Thomas
Newcomb*, and *Henry Hills*, deceas'd; Printers to
the Queens most Excellent Majesty. **1712.**

Queen Anne's Act for the restoration of Thames River salmon.

The original abundance of salmon was not restricted to the forested regions. Early in the eighteenth century, Daniel Defoe, in his *Tour Through the Whole Island of Great Britain* (1726), had described prosperous salmon fisheries in the north of Scotland where there were "salmon in such plenty as is scarce credible and so cheap that, to those who have any substance to buy with, it is not worth their while to catch it themselves" (Netboy 1974, 57). But even in Defoe's day the state of British salmon rivers was not uniform. Some runs were already declining.

Attempts to recover and restore English salmon stocks date at least as far back as a 1712 act passed during the reign of Queen Anne that provided for regulating fishing on the Thames River with the intent of reducing overfishing, so that "salmon fish, which are become very scarce by destroying great quantities of salmon . . . when they are out of season, or spawning, may become very plentiful and common in the said fishery, as they were formerly" (Anne Regina, 1712, 539). Shortly after Anne's death two years later, George I enacted a law to prevent blocking salmon from their spawning grounds in seventeen English rivers. The law renewed stiff fines for violations, and held that offending nets or weirs were to be destroyed at the owner's expense. In addition, salmon caught out of the proscribed fishing season were to be confiscated and destroyed.

Amid growing concern over the state of salmon runs, many British rivers still supported substantial salmon fisheries. Before 1800, even the smallest Scottish streams supported salmon fishing. In some places, salmon conservation efforts led to rather novel practices: At a dam built to power a cotton mill on the Ribble River, a bailiff conscientiously scooped up salmon and carried them over the dam once the fishway fell into disrepair. That most English salmon rivers remained well stocked at the close of the eighteenth century testifies to the effectiveness of early laws, as well as to centuries of concern over the effects of overfishing and small dams on salmon. Yet by the end of the nineteenth century salmon were extinct in many English rivers, casualties of the Industrial Revolution.

Before the advent of steam power, factories were located along swift-flowing streams so as to harness the energy of running water. The first modern factory, a water-powered silk mill, was built around 1720 along a salmon stream in Derby, in northern England. As great factories rose to replace little mills across the countryside it became customary to discharge industrial wastes into the nearest stream or river. Untreated sewage from the bleak towns that sprang up around the factories was dealt with in the same manner. Bacteria feasting on raw sewage from squalid industrializing cities turned rivers into oxygen-free dead zones deserted by salmon. Bars of partially oxidized sewage blocked the mouths of some tidal rivers. As the city of Manchester grew into an industrial wasteland, the river flowing through its center was transformed from a salmon stream into an open sewer. Many English rivers poisoned by pollution during the Industrial Revolution still have no salmon.

The Thames River salmon protected by Queen Anne were among the first to disappear altogether. Reliable early fishing records are rare, but in researching the Thames River salmon the aptly named Anthony Netboy reported that the holder of fishing rights at Maidenhead recorded the demise of the Thames salmon fishery in his diary (1974, 70). In 1801 this man caught 66 salmon, in 1812 he landed 18 salmon, and in 1821 he could catch only 2 salmon. He did not consider it worth fishing for salmon after that. Conflicting versions of when the last Thames salmon was caught include the story that George IV ate it after purchasing it for a guinea a pound.

The story of the Thames salmon played out all over the British Isles as salmon flowed from the countryside into the cities, as noted in the Quarterly Review: "[T]he natives of the countries though which salmon-rivers flow [have] become accustomed to see them taken and cased up for the great city, by scores and hundreds, without having it in their power to purchase a pound for their table" (1828, 529).

Salmon no longer fed the masses. Within a few decades salmon became an expensive luxury, as was noted in 1854 in the *Scotch Reformer's Gazette*: "In former times, salmon was a staple article of food in this country. . . . So different is the case now, that even with persons of comfortable means the salmon is but a rare luxury. It finds its way only to the tables of the well conditioned and affluent" (Young 1854, 82–83). The king of fish had become the fish of kings.

Pollution and royal appetites were not the only reasons for the salmon crisis in nineteenth-century England. Changes in the laws governing closed fishing seasons contributed to the decline. In 1828, George IV repealed the centuries-old act of James I and replaced it with the Act for the Preservation of the Salmon Fisheries in Scotland, which extended the season in which salmon could be taken from mid-August to mid-September, into the peak of the spawning runs.

The 1828 act also made the start of the closed season the same for the river and coastal fishery and dropped the prohibition against nets permanently affixed to platforms or pilings. Much of the ancient legislation to protect the salmon focused on preventing channel-spanning obstructions and nets in "the run of the fish." Such prohibitions were considered not to apply once the fish left the river for the ocean. Coastal fishing became a major industry now that fixed nets could be staked out along pilings extending from the shore and across the path of migrating salmon as they hugged the coast before heading out to sea. Large salmon-netting operations quickly sprang up along the coast and at the mouths of Scottish rivers. Simultaneously, the definition of where the river emptied into the sea began to creep upstream.

Decades later Andrew Young, the manager of the Duke of Sutherland's fishery, noted how salmon "must finish their course among the meshes of these destructive engines: they have no means to escape, for the one placed after the other, and always in the principal track that the fish run in, is certain destruction, and very few escape" (1854, 14).

The misnamed Act for the Preservation of the Salmon Fisheries backfired. Salmon runs crashed under the new regulations. At first, only naturalists and tenants who ran the fisheries noticed the decline of Scottish salmon in the aftermath of the 1828 act. But within five years of the 1828 act the devastating effects of changing Scotland's ancient fisheries laws started becoming apparent. Rampant overfishing was killing the goose that laid the golden eggs.

Declining salmon abundance drove fishermen out of business even as the rent of salmon fisheries dropped precipitously. The rent of a fishing station on the Ness River in Scotland fell from £211 in 1796 to £160 in 1815 and just £40 by 1833.

"The salmon fishing is now a complete failure, and is falling into the hands of the proprietors for want of tenants," commented Alexander Fraser. "This also is owing to too many traps laid for the capture of the fish. . . . [N]o breeders are left, and we may as well expect to see and hear the cuckoo amongst us at Christmas as that any fresh breeders should enter our rivers" (1833, 125–26). Fraser blamed changes in the law for decimating salmon runs. He understood that the combined effects of overfishing and many small dams that blocked salmon migration threatened to destroy the salmon fishery.

A pragmatist, Fraser proposed a simple way to increase the number of salmon in Scottish rivers: stop killing salmon six weeks before the end of the run to let enough breeders make it to their spawning grounds. He also proposed that salmon fishing be banned on Sundays, and that canal gates and mill sluices be opened at least twice a week from February to June to allow for passage of both juvenile salmon and any surviving spawned-out adults back down to the sea. His common-sense proposals for restoring the salmon fisheries were to avoid blocking the ability of salmon to migrate up- or downstream and to limit fishing intensity so as to not take the majority of the spawners.

Little changed, however. As the river fisheries continued to decline and rents were driven down, their proprietors began to worry. Eventually the general public became concerned, and by the early 1850s even popular periodicals such as *Bell's Life* were railing about the sad state of Scottish fisheries: "[N]o salmon river now, even the very best, is worth one-fourth of the value it possessed thirty years ago, and very many of them that were productive at that time must remain henceforth tenantless, as now they will not pay the outlay for watching and working them " (Young 1854, 95).

Overfishing was generally blamed for the rapid decline of the salmon fisheries. In the 1730s the number of salmon taken each year from the largest and most seaward salmon-fishing operation on the River Tweed was between 1,500 and 2,000 salmon. Several decades later, in the late eighteenth century, the operation was hauling in between 10,000 and 20,000 salmon. By 1818, however, despite increasing fishing effort, the catch dropped to a few thousand fish and by 1850 fell to well under a thousand salmon. Similarly, the salmon catch above the Perth Bridge on the River Tay by 1850 had declined to 15 percent of the 1790s levels.

Older salmon were becoming particularly rare. On the River Tay, the catch of grilse accounted for just a third of the total catch before 1800. Fifty years later the catch of grilse outnumbered that of adult salmon two to one. Records for the River Tweed show that by 1850, 85 percent of the catch were salmon making their first return from the sea. The supply of older fish was running out and an increasing proportion of the catch were fish that had not yet even spawned.

A print dating from 1846 of an adult salmon and a salmon grilse.

Despite the obvious crisis, commercial fishing interests generally opposed attempts to restrict fishing. For example, the writer Alexander
Russel noted in 1864 in *The Salmon*, "[A]lmost everywhere in the
United Kingdom, and especially on the 'common fisheries' of England
and Ireland, [fishermen] are more or less fully possessed with the notion that restrictions as to periods and modes of killing are invasions
and injustice, and that the more fish that are killed, the more will remain to be killed" (90).

Restraint no longer guided fishery management. In fact, the decline
of English salmon had the perverse effect of increasing fishing pressure
because of the increased value of the remaining fish. As salmon became
more valuable, regulating the salmon fishery became more difficult. As
salmon runs declined, commercial salmon fishing accelerated, poaching
flourished, and fishing laws became widely disregarded. By the 1860s
poaching had become a regular profession that people relied upon for a
weekly income. As poaching became more profitable, poachers became
bolder. In 1869, the inspector of salmon fisheries reported that during
an inspection of the River Tyne, as night fell he could see lights from
hundreds of boats engaged in illegal fishing. Salmon were a lucrative target now that a couple of fish were worth as much as a sheep or a pig.

In the few places where overfishing was not rife and habitat was in
good shape and not blocked by dams, salmon runs began to rebound

where sufficient river keepers were employed to keep poachers from spearing salmon on their spawning beds. But in most areas enforcement of fisheries laws remained a key problem and overfishing was rampant. The government appeared to have little interest in interrupting the activities of poachers and entrenched fishing interests.

It had even less interest in protecting the habitat of the fish. By 1868, all seventeen rivers protected by George I were either blocked by dams or poisoned by pollution. Only a little more than a third of the area drained by these rivers remained accessible to and usable by salmon. Over the course of a century and a half, salmon were locked out of almost two thirds of the length of these rivers. The following year, the inspector of salmon fisheries reported that salmon had been physically blocked from well over half of their potential range in the six largest English rivers.

Many with knowledge of the salmon fisheries were frustrated by governmental inaction. Some bemoaned the government's preoccupation with politics and its general apathy toward the state of salmon runs. Others, such as the fisheries manager Young, hinted more directly at the role of vested interests in preventing reforms: "[N]othing could arouse the Government and the Legislature of the country to even attempt to stem the current of destruction. . . for it seemed as if those that had the undoubted right to protect the fish had conspired together for their actual extermination" (1854, 16). Official inaction, apathy, and complicity contributed to degrading the British salmon fisheries.

As the salmon crisis deepened, public sentiment turned against overexploitation of the fisheries and momentum built for the government to restore salmon runs. Proposals to curtail overfishing included those for a several year ban on salmon fishing and the equally radical idea of consistently leaving half of the fish in the river. The causes of overfishing were as clear as the consequences of failure to address the problem. A writer in the *Quarterly Review* lamented:

> If in our greed we still continue to overfish, after the numerous warnings we have had, we must take the consequences in the probable extermination of the salmon. (1863, 422)

Many bills were introduced to reform salmon-fishing laws, but Parliament would not tackle the issue. In 1864 a long-time MP, Sir Robert Peel, remarked that he had never known a session of Parliament without a salmon bill. Despite the widespread conviction that something needed to be done, and the repeated introduction of bills to Parliament, nothing was done until the 1860s—other than the 1828 Act that *extended* the fishing season.

Politics stymied reform. In a letter to the *Inverness Courier*, Robert Wallace, a former member of Parliament, recalled with frustration the political paralysis over the salmon crisis:

> During the many years I was in Parliament, bills for the protection and increase of salmon in Scotland were unsuccessfully introduced in the House of Commons almost every session, until it became quite evident that selfish motives or gross ignorance, or both, prevailed so strongly that no rational alteration of the present destructive statutes could be obtained. For when we passed a wholesome change in the law through the House of Commons, it was certain of being either neglected or strangled in the House of Lords. (Young 1854, 100)

Wallace went on to describe how the owners of the river fisheries generally chose not to blame overfishing for the sad state of their business and instead preferred to blame mill-dam owners for declining salmon stocks. Wallace was convinced that because of such genuine conflicting interests it was the government's duty and responsibility to protect salmon runs to provide for a perpetual food source that served the common good. Even though it was clear to those without a vested interest that both dams and overfishing were gradually destroying the salmon fishery, it took three decades to muster enough political will to modify the 1828 act.

Although most attention focused on the impacts of overfishing and dams on salmon, habitat quality and condition were also recognized as important issues affecting the English salmon, particularly the straightening and clearing of rivers into navigable channels. Significant effort was expended to clear English rivers and improve inland navigation between the 1600s and early 1800s. Many rivers were straightened, dredged, and locked into place by levees. In his 1869 report, the inspector of salmon fisheries discussed how converting rivers into navigable waterways destroyed salmon spawning beds and significantly reduced the salmon-producing capacity of a river.

It was observed that increased land clearing and drainage in upland areas influenced flooding and siltation in downstream rivers and streams where salmon spawned: "[It] is not the drainage of the land having immediate frontage to the river that has produced these results," wrote Russel, "but the drainage of . . . the land in high-lying districts at the sources of the rivulets . . . in almost none of which are there any salmon, and in none any ownership of salmon-fisheries" (1864, 115–16). Land-use changes in a river basin's headwaters were understood to affect salmon production in downstream rivers. The effects of landscape-scale changes sweeping across the country on increasing stream turbidity and scouring away developing salmon embryos were discussed regularly in the Victorian press and periodicals.

Though the effects of stream pollution on salmon were known long before the Industrial Revolution, the government did little to discourage the befouling of streams and rivers with industrial or municipal wastes. Quite the contrary, English society relied on streams and rivers to remove untreated industrial and human waste from cities and towns and transport it to the sea. One by one, salmon runs were wiped out as English streams were blocked to provide power for industries that then poisoned the waters with their refuse.

By the middle of the nineteenth century the plight of English salmon aroused widespread public concern over the danger of extinction. In 1854, the year that Governor Isaac Stevens spent signing treaties in Washington Territory, the fisheries manager and writer Andrew Young warned that British fisheries laws were insufficient to protect salmon: "[T]he race of salmon . . . has been destroyed in more than half of these rivers, and the other half are fast following" (1854, 9). Young railed both at the government's laxity in passing or enforcing laws to protect salmon and at irresponsible proprietors who overexploited their fisheries.

But Young saved special scorn for a new breed of accomplice in the destruction of salmon runs. He loathed the kind of lawyer who searched riverbanks for seashells to help clients redefine where the rivers turned into sea and thereby allow them to stake their nets closer to spawning grounds: "Lawyers made many proprietors of fisheries believe that the Scotch Acts meant this, that, and the other thing; that, in these Acts, black was not black, nor white white, and that the fixtures [fish traps] in those days are not the fixtures in our days" (10).

Along with lawyers, Young blasted expert witnesses in fishing cases, having "invariably found them on the side of extermination." These new hired guns of the legal arena lined up against advocates of salmon conservation.

But the salmon also had new allies. Public personalities like Charles Dickens whipped up sentiment to do something about the perilous state of the salmon fisheries. Anthony Netboy reports how Dickens, writing in the July 20, 1861 issue of his weekly magazine *All Year Round,* rallied public attention to the cause of salmon conservation:

> The cry of "Salmon in Danger!" is now resounding throughout the length and breadth of the land. A few years, a little more over-population, a few more tons of factory poisons, a few fresh poaching devices . . . and the salmon will be gone—he will be extinct.
>
> Shall we not step in between wanton destruction . . . and so ward off the obloquy which will be attached to our age when the historians of the nineteen-sixties will be forced to record that: "The inhabitants of the last century destroyed the salmon." (Netboy, 1968, 186)

Public pressure led to government studies, and then more studies. In 1860, a Royal Commission of Inquiry was formed to investigate the state of the salmon fisheries. The commission found that salmon fisheries around the country were depressed as result of the long-recognized problems of legal overfishing, poaching, and proliferation of obstructions to fish passage together with the newer problems of pollution from mine and industrial wastes. The commission further recognized that efforts to conserve salmon were hindered by lack of an organized system of river and fishery management and by confusion and uncertainty as to the provisions of the law and the difficulty of enforcing penalties against offenders.

The following year, the Salmon Fisheries Act of 1861 standardized laws pertaining to salmon fisheries in England and Wales. A uniform season closed to fishing was adopted to combat the fraud and poaching that flourished under the umbrella of a complex web of overlapping seasons set by local authorities. The act also authorized both the appointment of inspectors of the salmon fisheries and regulation of the discharge of industrial waste into river and streams, specifically prohibiting "putting into any waters containing salmon any liquid or solid matter to such an extent as to cause the waters to poison or kill fish." The antipollution measures, however, were ineffective. Enforcement was relegated to notoriously lax local justices of the peace and no funds were provided to support their new activities.

Another key facet of the 1861 act was directed at fishing weirs and mill dams. It grandfathered existing illegal weirs while requiring future weirs to have a gap at least as wide as one tenth of the river's width to allow unimpeded fish passage. Unfortunately, by legalizing fish passage barriers constructed in violation of previous statutes dating back hundreds of years, the 1861 act began to lower the baseline against which to evaluate degradation of salmon streams.

Yet it would not have cost all that much to retrofit small dams to allow for fish passage. In the 1820s a mill owner, James Smith, invented a "salmon stair" to allow salmon to pass the dams at his mills on the River Teith in central Scotland. Like other Scottish mill owners, Smith took great interest in watching salmon trying to jump over his dam. Sympathetic to their plight, he designed a series of small steps separated by pools that was cheap and easy to build and allowed passage to salmon over formerly impassable dams. Installation of fish passways remained voluntary, however, and few dam owners chose to make the effort. Fifty years after Smith installed a salmon stair at his mill, the *Edinburgh Review* reported that English salmon were blocked from more than two thirds of their natural breeding-grounds, and that mill owners "resisted undertaking such work and small expenditure on the old plea of ruin and confiscation."

Another royal commission brought to the public's attention the extent of pollution and unsanitary conditions in rivers that were used as sewers in the industrial regions of the country. This led to the Rivers Pollution Act of 1876, which prohibited discharging sewage into streams and rivers. Enforcement proved the Achilles heel of the 1876 act, as polluters had to be beaten in costly court proceedings in order to curtail illegal discharges. Even then, penalties for failure to comply with court orders could be imposed only after further proceedings. Anthony Netboy recounted how a weir at Settle, rebuilt between 1860 and 1867 as an impassable 6-foot high embankment, was declared illegal but no action was taken to remove or modify it. Furthermore, dischargers of industrial wastes were held to less-stringent requirements than the general public. The Salmon Fisheries Reform Acts of 1861 and 1865, and the Rivers Pollution Act of 1876 slowed but did not reverse the decline of English rivers and their salmon fisheries.

Fifty years later, the Salmon and Freshwater Fisheries Act of 1923 consolidated and updated previous laws. Once again, new dams and alterations to existing dams were required to include fish ladders, now to be approved by the Ministry of Agriculture and Fisheries. As before, existing dams were exempt, thereby legitimizing dams constructed illegally after 1861. The 1923 act again banned the polluting of streams that would "make them harmful to fish." But just as before, impacts in place at the time of the act's passage were exempted as long as the "best practical means within reasonable cost were used to prevent such discharges doing injury to fisheries." Though the ambiguous definition of reasonable cost provided an obvious loophole, the 1923 act ended up proving a turning point in addressing river pollution, no doubt due to increasing public support for cleaning up polluted waterways.

Authority over river management was gradually centralized as efforts to clean up English rivers evolved over the next forty years. The River

Boards Act of 1948 established a single authority that would be an umbrella of boards responsible for fisheries, land drainage, and pollution prevention. These new boards were granted the power to impose levies to fund their work. Many fisheries authorities could now for the first time afford personnel to adequately monitor and enforce regulations. The Rivers Pollution Act of 1951 made it an offense (yet again!) to cause or knowingly permit pollution of a stream or river. Although fines were small, the river boards made substantial progress in combating stream pollution. As the tide began to turn against polluters, conflicts arose and Parliament transferred the powers of the river boards to river authorities that also had jurisdiction over water withdrawals for agricultural use. Local authorities were given power to appoint half of the members of the river authorities, the other half being appointed by government ministries. A century after passage of the reformist salmon fisheries acts of the 1860s and 1870s sought to restore British salmon, the annual catch in English and Welsh rivers had dropped to just a quarter of the 1870 catch.

The situation in Scotland was different. There, salmon fisheries remained stable through much of the twentieth century, as regulations restricted fishing gear to relatively inefficient methods and aggressive efforts were undertaken by fishery boards to clean up rivers and remove or modify barriers to fish passage. Still, increased ocean mortality in the late twentieth century further depressed of many stocks. The specific cause remains something of a mystery. Some blame open-ocean fishing or drift-netting in coastal waters. Others blame increasing marine pollution for killing off the foundation of the marine food chain on which the salmon depend.

By 1985 it was clear that the British government lacked the political will to address overfishing in their coastal waters. A fresh approach to salmon recovery was needed, one that didn't rely on government regulation. As often happens, the new approach was based on an old one.

Unlike England's public salmon fisheries, Scottish salmon fisheries had been private property for centuries, so they could be bought. In 1872, the marquis of Huntley established an association to buy and retire salmon netting operations on the River Dee, which flows into the

North Sea near Aberdeen. Huntley owned 30 miles of waterfront in the upper portions of the basin. Under Huntley's leadership, the upstream fishery owners gradually purchased the fishing rights in the lower river, eventually closing down most fishing downstream of their holdings— thus ensuring that fish would not be impeded in reaching their upstream spawning grounds.

Taking a cue from Huntley's efforts a century earlier, in the late 1980s the Atlantic Salmon Conservation Trust began to buy salmon fishing rights with the plan of phasing out fishing efforts until stocks recovered. Within a year the trust had purchased more than two hundred fishing stations that together accounted for about a quarter of the harvest of Scottish salmon. Runs began to rebound where the nets were closed down, and the conservation approach based on purchase of fishing rights quickly spread across Scotland. Private interests were able to curtail overfishing in Scotland, something that governments generally have lacked the will to do where the prevailing view was that the fishery was a public good to be exploited by all. Eventually, the value of Scottish recreational fisheries came to exceed the value of the commercial fishery.

The story of the English and Scottish salmon has a number of lessons for our modern salmon crisis. The process of destroying a fishery by overfishing takes so long that individuals have little incentive to stop catching the fish, even if they are aware of the impending train wreck. Hence, nonmarket incentives are needed to restrain unfettered fishing. But local control of regulation and enforcement proved ineffective once salmon became valuable enough to undercut general compliance with regulations. Only after cooperative boards with both local and governmental representation were set up did salmon protection become both a priority and an effective practice. But even so, landscape changes that restrict or reshape salmon habitat can slowly accumulate into huge negative impacts on salmon populations, especially if new regulations grandfather in existing violations of previous laws and regulations.

Iceland's salmon fishery provides a contrasting example of exemplary modern salmon conservation. Commercial fishing at sea is prohibited.

Although salmon are regarded as the property of landowners along rivers and streams, fishing is subject to strict regulations. Salmon traps are prohibited. During the three month fishing season, netting is permitted for only half of each week and rod fishing is allowed for just twelve hours each day. The total catch of Icelandic salmon from the 1930s to 1970s is estimated to have been at least a third and no more than about half the fish returning to Iceland's rivers. Managed on a river-by-river basis in a partnership between the Ministry of Agriculture and local fishing associations, Iceland's salmon runs actually increased during the second half of the twentieth century.

In the early 1990s my wife, Anne, and I took a river taxi up the Thames River from London to Hampstead Palace, home of the notorious King Henry VIII (who unlike some of England's other Henrys seems to have made no effort to protect salmon). For most of the trip we chugged along through urban blight and factories, the Thames as it flows through London being less like a river than like a large concrete-walled canal. We couldn't imagine that the Thames had once supported thriving runs of salmon along its lower 65 miles. It seemed inconceivable that Caesar's legions caught salmon in downtown London. Even more astounding is that a Thames salmon holds the size record for a British River, an 83-pound beast caught in 1821. Soon after that champion fish was landed, the Thames salmon were gone.

For over a century no salmon roamed the river, but restoration and pollution control efforts in the last several decades are starting to pay off. Salmon are starting to return to the Thames. Just as in the rivers affected by the eruption of Mount St. Helens, salmon can stage a comeback if given a chance. They may have been exterminated throughout most of their original range, and many rivers that formerly hosted large salmon runs remain barren, but European salmon are not yet extinct. Efforts are underway to rebuild continental Europe's stocks from the few remnant runs. Restrictions on fishing are helping to make the replenishment of some European salmon stocks a real possibility, and a key focus of current efforts in England is to open rivers up to salmon again by removing obstacles such as dams and nets.

Techniques and experiences of the Old World salmon fisheries moved across the ocean with immigrants to the New World, but the transfer of experience in salmon fishing from Europe to North America was uneven. Many lessons about salmon management learned slowly in Europe were ignored entirely in North America. Some have slowly been learned all over again.

New World Salmon

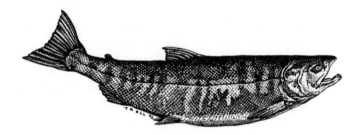

*Dim visions we still get of miraculous draughts of fishes, and
heaps uncountable by the riverside, from tales of our seniors.*

Henry David Thoreau, *A Week on the
Concord and Merrimak Rivers*, 1849

T HE PACIFIC NORTHWEST WAS ONE OF THE LAST BLANK
spaces on Europeans' world map, a mystery shrouded in clouds
when Queen Anne's geographers, accepting Sir Francis Drake's interpretation of Puget Sound as the northern end of the Gulf of California,
portrayed California as an island. Some theorize that Drake's map was
part of a sixteenth-century disinformation campaign intended to mislead
rival Spanish explorers beginning to sail north from Central America.

Whether by mistake or deceit, the basic geography of the region remained unknown for centuries. On some maps Puget Sound was represented as a great inland sea, and the location of the Great River of the

West—the Columbia River—depended mostly on the mapmaker's imagination. A veil of fog, clouds, and perpetual rain obscured the geography of the Pacific Northwest until George Vancouver mapped the region's coastline in 1793. A moment later in history, Lewis and Clark marched over the Continental Divide and floated down the Columbia River to the Pacific Ocean. Their expedition filled in the geography of the western interior and finally connected the maps of eastern and western North America.

I keep forgetting that the Seattle I know is brand-new, a recent addition to the landscape. Though the transformation from impenetrable forest to modern city happened in a geological instant, the dramatic changes that accumulate from daily experiences seem imperceptibly slow by human standards. The primeval forest that blanketed Seattle for 98 percent of the last five thousand years disappeared in a little over a century. Even so, I occasionally find myself thinking of the huge trees and logjam-filled rivers of Olympic National Park as being unusual rather than as what they really are: the natural norm for western Washington. It is difficult to imagine the original forest, let alone find evidence of it now.

Still, every now and then I am reminded that I live in a recently cleared forest. Digging in our front yard to replace warped front steps my wife and I found that the culprit behind the settling was a void left by the rotting stump of a huge tree buried in 1918 when the house was built on what was then the edge of the city. Today's landscape presents a strange new world for salmon that evolved beneath the region's old-growth forests.

Hard as it is to imagine that what one sees every day is not normal, some experiences show us how unusual our modern world is in evolutionary terms. Several years ago I flew over the Amazon on an overnight flight returning home from a research trip in the river's Bolivian headwaters. We had floated down 600 miles of the Rio Beni, a tributary of the Amazon. We started our journey where the river leaves the Andes and stopped every so often to drill cores into the floodplain and sample sediments that would be analyzed for the presence of both mercury (introduced by upstream gold mining) and a lead isotope that can be used to date the time of sediment deposition. Seven of us traveled in a large modified canoe for almost two weeks down the river, as large as the Mississippi but just a small tributary of the Amazon. Throughout the trip I was struck by how odd it felt to move through a continuous tract of virgin forest, broken in only a few places where small villages clung to the riverbanks. Later, on the flight home, the forest appeared endless as I stared out a plane window down into the jungle. The sky was clear and the southern stars riotous. But my attention was focused on the void below as we flew hour after hour over nothing but blackness.

Here was a taste of what the early settlers in New England or the Pacific Northwest must have felt when they arrived and found an endless wilderness of trees. Colonial descriptions of New England's forest echo my own feelings from the upper Amazon. John Bartram, writing in 1751, described the canopy of the eastern forests: "We observed the tops of the trees to be so close to one another for many miles together, that there is no seeing which way the clouds drive, nor which way the wind sets: and it seems almost as if the sun had never shown on the ground, since the creation" (37).

When one is walking around Boston it is difficult to reconcile Bartram's description of the original forest with the Atlantic seaboard of today. It is hard enough to conjure up the native forest while standing on the waterfront in downtown Seattle where one can still see the last remnant old-growth on the horizon. It is harder still to imagine the primeval English landscape, whether walking around London or the surrounding countryside.

More forest now covers New England than at any time in the last few centuries. The nineteenth-century abandonment of farming as population surged westward allowed extensive regrowth of forest cover in the rural east. In many places, the eastern forest today presents a more continuous cover than the managed forests of the Pacific Northwest.

In the mid-1990s I was amazed by the eastern forest when I flew to rural Pennsylvania to give a lecture at Penn State. During the descent I was astounded by ridgetops without angular clear cuts, strange bald patches, or bare earth. Instead, an almost unbroken canopy blanketed the hills, a striking contrast to the geometric maze of clearcuts one sees flying over the forests of the West. Much to my amazement, I found

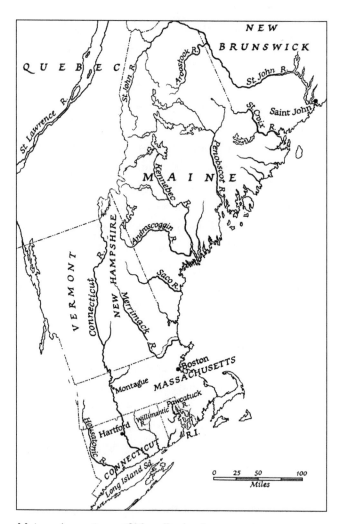

Major salmon rivers of New England.

myself thinking that New England's forest was now more like the Amazon than the Pacific Northwest. Except of course for the color: in the fall, a sea of red, leaves on fire with autumn glow.

The original range of the Atlantic salmon on the Eastern Seaboard extended from Long Island Sound over a thousand miles north almost to Hudson's Bay, and inland to impassable Niagara Falls. Prior to European colonization most streams and rivers from the Connecticut River to the waterways of northern Labrador swelled each year with salmon.

Minimum estimates of pre-contact salmon abundance on the Eastern Seaboard range from 5 million to 12 million fish. Some authorities describe the pre-contact salmon runs of the eastern United States as comparable to those of the Pacific Northwest.

Native Americans within the northern portion of the range of the Atlantic salmon—essentially modern Canada—maintained seasonal settlements they occupied in the summer for fishing. During the winter months they lived as nomadic hunters. To the south, in New England, agriculture was more prevalent and the native population maintained well-defined villages. Whether the people were nomadic or more sedentary, their settlements along rivers and lakes were located to exploit salmon fisheries throughout the range of the Atlantic salmon.

The native diet on the Eastern Seaboard was quite diversified, including abundant eels, catfish, crab, and terrestrial game. This may explain why the practice of drying and hoarding salmon to provide an entire year's sustenance does not appear to have been as prevalent in the East as in the Pacific Northwest. Salmon nonetheless played an important role in native societies. Salmon were even the primary food source in some areas of the Eastern Seaboard. In 1583 a European visitor to Newfoundland, Stephen Parmenius, wrote home, "For the most part their [the natives'] food is fish rather than anything else, and especially salmon, of which they have great abundance; and although there are many kinds of birds and fruits there, yet they make no account of anything but fishes" (Dunfield 1985, 14).

Other early observers also reported that salmon were an important staple of native diets in the Northeast. New Brunswick's Restigouche people adopted the salmon as their tribal symbol, adorning their canoes, clothing, and bodies with images of salmon. They are said to have been able to immediately tell, just from its appearance, which river a fish came from. Newfoundland's Beothuck people traditionally buried dried or smoked salmon along with their dead. At the time of first contact with European explorers, the native population of the Gulf of St. Lawrence depended heavily on salmon. Artifacts dating to 4000 B.C. indicate reliance on salmon in some areas of the Eastern Seaboard for thousands of years.

The Native American population on the Eastern Seaboard of Canada has been estimated as less than ten thousand at the time of early French exploration. In 1610, the population of present-day Nova Scotia was estimated to be about one person for every fifteen square miles. The impact of such a low human population density on salmon could not have been very great, especially with the diverse resource base and food sources available to native peoples and early settlers.

Like their neighbors to the north, Native Americans of present-day New England knew where to catch salmon, setting up camps where the fish congregated at the head of coastal streams or below barriers to upstream migration. They knew how to catch them, employing traps, nets, spears, and even natural toxins to stun the fish. Though they used a range of fishing methods, a three-pronged spear was the primary fishing tool prior to European contact. Spearfishing, a take-it-with-you technology well suited for life on the move, fit the nomadic way of life of the Northeast Indians. Accounts from the late 1500s and 1600s also describe native use of fish weirs and artificially constructed fishponds. The low human population density and high salmon abundance (as in the Pacific Northwest, the ratio of salmon to people was about 1,000 to 1) leads most researchers to conclude that Native American salmon fishing had little impact on the resource as a whole. Clearly, it was sustainable to the degree that both salmon and salmon-eating people populated the region for millennia.

That changed after the arrival of Europeans. Visits to North America recorded in the Viking sagas in about A.D. 995 provide the first datable historical record of salmon in the New World. The Vikings were adept salmon fishermen who had gained experience in their native Scandinavia and colonies in Iceland and Greenland. One early saga noted abundant salmon in the rivers of Newfoundland, observing that the fish were larger than the Norse people had ever seen before. After the Vikings abandoned their North American settlements by 1014 little more was recorded about North American salmon for almost 500 years.

Within decades of Columbus's voyage, the untapped resources of the New World enticed the English, Portuguese, and French to exploit Newfoundland's fisheries. John Cabot, Gaspar Corte-Real, and Jacques

Cartier all reported finding abundant salmon along the coast of New-foundland in the late 1400s and early 1500s. Cod, the initial target of the New World fisheries, could be processed and packed in salt-laden barrels for curing and shipping back to Europe. As the industry expanded, temporary on-shore drying stations became established to facilitate processing the catch. Stations needed freshwater sources such as the outlets of coastal streams or rivers. Fishermen could hardly escape noticing the hordes of salmon in the waters flowing by their encampments. Salmon quickly joined the flow of cured fish moving across the Atlantic. By the late 1500s, the British claimed sovereignty over Newfoundland's rich fishing grounds and dispatched a military presence to enforce their claim.

Report after report documented abundant salmon in the rivers of northeastern North America. Robert Juet, an officer who accompanied Henry Hudson on his third voyage (1609), recorded observations in his diary about the fauna of present-day New York State. On September 14, 1609, as they started sailing up the Hudson River, Juet noted that the river was "full of fish." The following day as they sailed past the Highlands and on to the Catskills, Juet's diary records that he saw "great store of Salmons in the River" (Juet 1909, 21). Although there is no real evidence that salmon inhabited the Hudson, large runs of fish greeted European explorers in Connecticut's rivers.

Early reports from farther north, in Massachusetts, stressed the new colony's salmon runs. Captain John Smith reported in his description of New England that "much salmon some have found up the Rivers" (Smith 1865, 36). Just twelve years after the Pilgrims landed, Thomas Morton, a member of the colony, expressed his view of the promising commercial potential of New England's rivers: "Of salmon, there is great abundance, and these may be allowed for a commodity" (Pearson 1972, 188). Two years later, William Wood, another member of the colony, further highlighted the abundance of cheap salmon in *New Englands Prospect*: "I set not down the price of fish there, because it is so cheape. . . . The Sammon is as good as it is in England and in great plenty" (37). It was no accident that five of the first seven settlements in New England were located on salmon streams.

Accounts of plentiful salmon were not restricted to New England. Nicolas Denys, an early governor of the French territories in Canada known as Acadia, devoted considerable attention to salmon in describing the natural history of his territory. Denys's 1672 account of salmon fishing on what was probably the Guysborough River in present-day Nova Scotia provides a more down-to-earth description of the early abundance of salmon than the colorful "you could cross the river on their backs" stories common in American salmon country:

> I made a cast of the seine at its entrance, where it took so great a quantity of Salmon that ten men could not haul it to land, and although it was new, had it not broken the Salmon would have carried it off. We had still a boat full of them. The Salmon there are large; the smallest are three feet long. On another occasion I went to fish four leagues up the river, as high as boats could go. There are two pools into which I had the seine cast; in one I took enough Salmon Trout to fill a barrel, and in the other a hundred and twenty Salmon. (Denys 1908, 166)

That one pool yielded to Denys more than twice the number of salmon caught in the entire river three hundred years later. On another unnamed river Denys reported finding salmon 6 feet long. According to Atlantic salmon sport-catch statistics for 1972 no fish caught even approached 3 feet.

Yet even fishermen like Denys did not always appreciate the abundance of New World salmon. He compared the nuisance presented by the Miramichi River salmon to that of the now-extinct passenger pigeons: "If the Pigeons plagued us by their abundance, the Salmon gave us even more trouble. So large a quantity of them enters into this river that at night one is unable to sleep, so great is the noise they make in falling upon the water after having thrown or darted themselves in to the air" (Denys 1908, 199).

Though they sometimes considered the fish pests, the settlers' interest in exploiting the hordes of salmon that returned to the rivers each year grew rapidly throughout the colonies. Initially salmon for local consumption were caught in simple dip nets, but unrestricted access to the fisheries encouraged growth of commercial salmon fishing as the colonial economy developed. In 1628, eight years after the Pilgrims established their colony at Plymouth, a French visitor described an elaborate fish weir they had constructed on a stream. It consisted of a set of two wooden trellis-work dams that passed water but not fish. Into these structures they installed wooden doors that slid up and down to trap migrating salmon.

By 1685 New England's population had exploded to 50,000 people and squabbling over access to salmon fishing locations is recorded in disputes over boundaries between adjacent townships. Regulation of the New England fisheries at first concentrated on guaranteeing that each species of fish was packed separately and that the barrels contained the proper amount of fish—concern focused more on access to and the preservation of the commodity than of the resource.

Salmon were so numerous that they were put to other uses than just food. The colonists' reliance on plow-based agriculture to cultivate corn quickly exhausted New England's soil. As early as the 1630s the English colonists were already applying salmon to fertilize their fields. Although effective as a fertilizer, the intolerable smell of rotting fish attracted wild animals and repelled travelers. Faced with ruined soil, farmers along spawning rivers put up with the stench as long as there were plenty of fish to fortify their fields.

Common as New England's salmon were, they were seldom sold before 1700. Most families salted casks of salmon for their own use, but few bothered to sell them when salmon fetched less than a penny per pound. It simply wasn't worth the effort.

In the mid-1700s falling salmon populations began to be noticed in what is now modern Canada. On Prince Edward Island, old-timers claimed that formerly large salmon runs had been greatly reduced by the fishing enterprises of ships bound for the French market. Although disrupted by periodic wars between Britain and France, the Nova Scotia salmon fishery continued to grow. In Newfoundland, the British colonial catch of 1757 was almost seven and a half times that of the 1720s. Even so, Newfoundland still had lots of salmon. Captain William

Pote, captured by Indians in 1745, reported that his captors caught fifty-four salmon in the course of a few hours.

It was in New England that the impact of changes in land use first began to decimate the New World salmon. As the colonies prospered, an explosive growth in small dams to power mills began to block salmon from their spawning grounds. Concern over preservation of the river fisheries motivated the colonial legislature to enact in 1709 the first of a long series of laws to protect salmon and other river fish. The act forbade the construction of new milldams and other obstructions to fish passage. It also granted counties the power to regulate fishing for the public good as well as to declare fish-passage blockages a common nuisance and dismantle them. The law, however, exempted existing milldams, starting an ongoing pattern of ignoring existing impacts and only regulating future impacts.

By 1728, the supply of local salmon had dwindled enough that Mr. H. Witton of Boston wrote to his friend Anthony Morse that he would pay for a man and a horse to bring a salmon to him in the city. Although Witton desired three or more salmon to make the expense worthwhile, he was willing to take anything, advising his friend that if there were but a single salmon to be had he should send it to him forthwith. Just as in Old England, New England's growing cities began sucking salmon in from the surrounding countryside.

The effects of overfishing and river blockages that were the target of fisheries laws in the British Isles were beginning to be recognized in North America. Even in the Connecticut River, which in the early 1700s remained free from obstructions, salmon stocks began to shrink as large nets were stretched across or dragged through rivers, replacing hand-held spears and scoop nets. By the mid-1700s salmon were disappearing from streams and rivers south of Maine's Penobscot River owing to the industrious New Englanders' talent for fishing, land clearing, and dam building.

The decline of the salmon fishery in New Hampshire's Piscataqua River (literally, "fishwater river" but now known as Salmon Falls River) provides a striking example of the state of New England's salmon in the early 1700s. In 1717 it was reported that the river was so full of salmon that a thousand tons could be taken in a season. At 10 to 20 pounds per fish this would be somewhere between 100,000 to 200,000 fish. Even if this estimate is inflated, the river had an impressive salmon run. By 1750, however, the combined effects of overfishing and milldams had reduced salmon runs enough to lead colonist James Birket to note, "Formerly this river . . . was well stored with salmon which they took in plenty, but of late they have quite forsaken this river, occasioned, it is believed, from the number of saw mills" (Pearson 1972, 208).

Despite such warning signs of impending decline, salmon remained numerous in most of New England's major rivers. Through the 1760s wagonloads of salmon were taken annually from the Connecticut River at the falls at South Hadley. In this same period, a single large net on the Merrimack River yielded up to a hundred salmon a day at the peak of the run. Still, concerned colonial legislatures began introducing bills to regulate river fishing and allow for passage of spawning runs.

Salmon remained quite abundant to the north in less densely populated Labrador. The fisherman George Cartwright's descriptions of salmon fishing at Eagle River in 1775 echo those of earlier days farther south: "We have 140 tierces [42,000 pounds] ashore, but have had to take up two nets, as fish got in too fast. The big pool is so full of salmon, you could not fire a musket ball into it without injuring some" (Dunfield 1985, 56). In August the following year Cartwright's operation took over twelve hundred salmon from one pool in just five days. In his journal Cartwright boasted, "Few escape there. . . . My ten nets, each forty fathoms long, fastened end to end, stretch right across the stream. . . . In Eagle River we are killing 750 salmon a day . . . and we would have killed more had we had more nets" (Netboy 1968, 349–50). Restraint simply was not an issue. The goal was to take all one could. Nets like Cartwright's were reported to completely block salmon streams throughout Labrador by the end of the 1770s.

As salmon fishing in New England began to decline, Yankee fishermen ventured farther north. By 1775, the pressure of the New Englanders on the Newfoundland fishery, together with British annoyance over the colonist's rebellious attitude, led Parliament to forbid New Englanders from fishing loyalist waters to the north. The move enraged fishermen accustomed to taking advantage of the Newfoundland and Labrador fisheries. Their anger helped fuel rising public sentiment against the Crown.

By the time the British colonies boiled over into revolution, the North American salmon fishery was catching up to a million fish a year. This bounty was traded within the colonies and exported to England, France, and northern Europe. The American Revolution disrupted fishing in loyalist waters of Canada even as it increased pressure on the salmon in the breakaway colonies' waterways. The British navy controlled the seas during the Revolution, and river-based salmon fishing became increasingly important once the British blockade prevented external trade and curtailed ocean fishing. So the Continental Army was forced to eat salmon and other river fish.

As the Revolution dragged on it placed increasing pressure on the river fisheries to provide sustenance for the colonists and their army. In 1778, the Continental Congress signed a contract for 10,000 barrels of salmon and shad to feed the Army. Smoked salmon was shipped from Maine to supply the troops. Salmon from Lake Champlain supplied the Revolutionary army at Ticonderoga. William Gilliland alone provided the Continental Army with fifteen hundred salmon in one year from his fish weir in the wilderness around Lake Champlain. Effectively locked out of the North Atlantic for a decade, New England's fishermen let salmon run the blockade for them.

The human aftermath of the American Revolution displaced as many as 100,000 loyalists north to the remaining British colonies. Those who fled New England dramatically increased the population in New Brunswick and other parts of modern Canada. This in turn increased pressure on salmon north of the already stressed New England stocks. The human population at the mouth of the St. John River increased from less than 100 souls to over 14,000 people after the Treaty of Versailles ended the war in 1783. The population of Parrtown (St. John), New Brunswick, grew to support more inhabitants than the entire non-native population of the territory east of the Penobscot River before the Revolution.

The northern salmon presented familiar opportunities to the new arrivals. Land grants to the displaced colonists included rights to river fisheries bordering the grant. People moved quickly to acquire choice properties from which to catch salmon. The intensity of expanding unregulated fishing worried Benjamin Marston, the sheriff of Northumberland County. In 1785 he described how salmon fishing nets on the Miramichi River "so far extended into the river from each shore as in some places to interlock . . . [and] absolutely stop the whole body of [salmon] from getting up to their spawning places and must eventually, much lessen, if not destroy the breed" (Dunfield 1985, 65).

Salmon fishing expanded so quickly that the legislatures of both Nova Scotia (1786) and then New Brunswick (1810) moved to pass legislation that outlawed impeding the return of salmon to their spawning grounds. In the subsequent decades, additional regulations further restricted fishing. By 1810 milldams were required to provide for fish passage throughout the remaining British colonies. Regulatory authority for other aspects of river fisheries was left to local government, which as in New England tended to concentrate more on inspection and allocation of the commodity than protection of the resource.

Meanwhile, salmon were becoming big business. Exports from the remaining colonies of the British Crown reached more than 4 million pounds by 1814. The growing economic importance of the salmon fisheries gave fishing interests substantial clout in post-Revolutionary North America.

Access to the fisheries of the loyalist northern colonies was a key bargaining point at the end of the Revolutionary War. John Adams told his British negotiating counterparts that peace was not possible without a guarantee of American access to Newfoundland's fisheries. In their haste to sue for peace with their former colonies, the British gave United States citizens the right to fish in "bays and creeks of all . . . of His Britannic Majesty's dominions in America." Though it took twenty years to rebuild American fishing capabilities after the Revolution, by the early 1800s American vessels filled every bay and inlet in Labrador. Local fishermen petitioned the British government for protection from the aggressive New Englanders, claiming they threatened the future of commercial salmon fishing in their provinces.

Unlike the Revolutionary War, the War of 1812 did little to curtail American fishing in the remaining British colonies. American interests viewed the hostilities as irrelevant to their fishing rights. Yankee boats are estimated to have accounted for half the Newfoundland salmon catch in 1814. Although Newfoundlanders saw the war as a means to revoke American fishing privileges granted by the Treaty of Versailles, access to their neighbor's fisheries was again a make-or-break issue for American negotiators. The sensitive issue was left out of the 1814 peace treaty, leaving American rights unchanged. By 1829 over 90 percent of the fishing vessels operating in the Bay of Fundy were American-owned and -operated.

As salmon stocks began to dwindle, conflicts arose among fishermen. Particularly controversial were nets that blocked the entire width of a stream or river, preventing salmon from ascending. Provincial legislatures passed acts for the preservation of salmon which banned the use of nets, weirs, or other devices at or near the mouth of rivers and provided for penalties of both a fine and a month in jail for a first offense. But the new regulations had little effect. An evening's catch could more than cover the fine in the unlikely event that one was caught and convicted of violating the law. By 1815 salmon were becoming so scarce in New Brunswick that commercial efforts were barely worthwhile.

New England's salmon fishery continued to expand in the wake of the American Revolution. Viewing the resource as inexhaustible, Congress encouraged development of the fishing base of the new country by subsidizing fish exports. New Englanders happily obliged. After all, as Benjamin Franklin put it, salmon were like "bits of silver pulled out of the water."

In the 1780s and '90s, salmon still fed half the population along the Connecticut River. More than enough could be caught to satisfy local demand. Fish sellers required those wishing to purchase shad also to buy salmon. But about 1795, a dam built at South Hadley Falls near Hollyoke, Massachusetts, impeded fish passage, and a subsequent mill and dam built at Montague, Massachusetts, in 1798 completely eliminated upstream runs. Between 1787 and 1798 the price of Connecticut River salmon quadrupled, from two to eight pence per pound. Salmon were caught along the whole river until 1800, but by 1815 Connecticut River salmon were rare.

Similar stories played out on rivers throughout New England. In 1773 concern over declining salmon runs in Massachusetts' Merrimack River led to a General Court order barring the common practice of placing river-spanning nets. This and other such orders provided only a temporary reprieve. Salmon runs to the upper Merrimack River were sacrificed to a New Hampshire dam in 1812; the Lowell dam in 1822 blocked access to 80 percent of the salmon's range on the river; and then the runs in the rest of the river succumbed to the Lawrence Dam in Massachusetts in 1848. In 1805 a fisherman could expect to catch thirty salmon on a good day. By 1830 it was rare to catch ten salmon, and by the 1850s there were no salmon left to catch. As fish populations crashed throughout the region, New England's salmon virtually disappeared in the course of a few decades.

New threats to the salmon also developed as the European population in North America mushroomed. Immigrants flooding to the New World transformed the landscape as the ax and plow pushed inland. By the 1850s, New England's forests had been stripped from river valleys and three quarters of the region's arable land was under cultivation. Land clearing and deforestation triggered soil erosion, which filled streams and rivers with silt, which buried spawning gravel and led to increased water flows, which scoured streambeds. The landscape literally began to shift away from under New England's salmon.

The rise of the eastern timber industry in the nineteenth century finished off many salmon runs that had so far remained relatively unscathed. Dams for sawmills (milldams) proliferated across the landscape not only blocking salmon from spawning grounds but serving as convenient fishing locations for hungry loggers. Many runs were destroyed. Others were just decimated as log drives scoured out spawning gravel and piles of sawdust and mill waste up to 20 feet high choked some rivers.

James McGregor described the effects of the timber industry on New Brunswick's salmon runs in the early 1800s in his *Historical and Descriptive Sketches of the Maritime Colonies of British North Amer-*

ica. Even in streams not blocked by milldams the annual log drives destroyed incubating eggs and growing fry. Many rivers were used as roads, with horse teams dragging barges and scows through shallow areas and spawning gravels. On the Miramichi River, wrote McGregor:

> Those who first settled on the banks of the river were attracted thither by its plentiful salmon fishery, which formed for some years a profitable source of enterprize. The exportation of timber has since then superseded almost every other pursuit; and the waters of the river being much disturbed by vessels, boats, and rafts of timber, &c., an extraordinary decrease in quantity has followed in the salmon fishery. (1828, 162–63)

Similar stories described the state of other rivers. The destructive side effects of logging on salmon were not hard to see; they were common knowledge.

Measures to recover salmon stocks by restricting fishing intensity proved futile in the face of habitat loss and degradation. Legislation requiring milldams to provide fishways was generally ignored. By 1837, in the state of Maine were approximately 1,300 sawmills, 250 along the Penobscot River alone. The threat to salmon from such a proliferation of dams was no mystery. Contemporary observers reported that almost every stream large enough to power a sawmill was blocked at from one to over a dozen points. An average of 15 new timber mills were built each year in British North America between 1812 and the middle of the century, more than 500 in all. By 1850 over half of their original habitat in eastern North America was cut off to salmon, and half of that which remained accessible was degraded by pollution, sedimentation, or direct mechanical disturbance from log drives.

The lumber barons of the British colonies were politically connected and were buffered from public concern. Besides, as the fishing declined people turned to the timber industry for steady work. By about 1850 as many as three out of four people in some parts of British North America were employed by timber operations. Adopting the implicit policy that lumbering was more important than fishing, provincial governments chose not to enforce fish-passage regulations. Statutes enacted to preserve salmon focused on overfishing. Few measures were taken to deal with the disturbance of spawning beds and rivers from transporting logs.

Salmon fishing in British North America peaked in the 1840s and then declined rapidly. Salmon exports from Newfoundland dropped by almost half between 1842 and 1846, and from New Brunswick, by almost a third. Nova Scotia became a net importer of salmon in 1848. By the 1850s only Labrador retained a relatively undepleted commercial salmon fishery. Even sportsmen began to complain about the difficulty of finding salmon in Lower Canada.

South of the border, in New England, the spread of factories for cotton and textile manufacturing blocked many rivers. By 1840, most of New England's 800 factories used small dams to tap water power. Mirroring developments on the Merrimack, a dam constructed across the Kennebec River at Augusta in 1837 locked salmon out of more than 80 percent of the river. A later dam further squeezed spawning into just a half mile above the mouth of the river, driving Kennebec River salmon close to extinction by the 1860s.

Just as in England, salmon became more valuable as they became scarcer. Bostonians who remembered the taste of fresh salmon were still keen to acquire them. Many in the prosperous city could afford the escalating price. The first of the Kennebec salmon caught in 1840 sold in Boston for an astounding forty-two dollars. Restricted to but a fraction of their prior range, yet highly desired for the table, salmon became ever scarcer in rivers throughout New England. Imports from north of the border tripled between 1832 and 1848. Though the supply of salmon was nearly exhausted, demand remained high, as did the price: about a dollar a pound.

By 1846 New England's salmon were commercially extinct outside of a few rivers in Maine, and one observer noted in 1855: "In former years, salmon frequented every river eastward of the Hudson, but lately, dams, saw-mills, and other obstructions, the result of Yankee enterprise, have driven them from the United States" (Hardy 1855, vol. 2, 105). The Kennebec River was now effectively the southern limit of the range Atlantic salmon. Shifting the southern boundary of the salmon's range a thousand miles north left 25,000 square miles of coastal territory with empty rivers that had once held salmon.

Writing in *A Week on the Concord and Merrimack Rivers*, Henry David Thoreau lamented the absence of salmon from the Concord River:

> Salmon . . . were formerly abundant here, and taken in weirs by the Indians, who taught this method to the whites, by whom they were used as food and as manure, until the dam, and afterward the canal at Billerica, and the factories at Lowell, put an end to their migrations. . . . Perchance, after a few thousands of years, if the fishes will be patient, and pass their summers elsewhere, meanwhile, nature will have leveled the Billerica dam, and the Lowell factories, and the Grass-ground River run clear again, to be explored by new migratory shoals. (Thoreau 1849, 36)

Not only was the range of salmon shrinking in the rivers of North America, but the salmon themselves were shrinking. From 1830 to 1850 the number of Restigouche River salmon required to fill a 200-pound barrel rose from roughly ten to twenty fish. As overfishing reduced salmon numbers, it preferentially removed the larger fish and older age classes. The net result was smaller salmon capable only of digging shallower nests that were more susceptible to being scoured out by floods.

Once the virgin forest had been cut and the loggers had moved on for points west, the continuing decline of the salmon fisheries received renewed attention. Beginning with New Brunswick in 1846, the British provinces commissioned official investigations into the state of the salmon fisheries. Throughout the provinces such reports described salmon runs decimated by impassable dams, degradation and disturbance of spawning grounds, and overfishing unaffected by ineffectual and unenforced laws. Investigators consistently found few or no salmon in rivers where they were formerly abundant. Report after report expressed an urgent need for reforms.

Although there were laws designed to protect the resource, civil authorities were generally unwilling or unable to enforce them. Consequently, the fishing populace blatantly abused or circumvented the rules, and there was no pretense of enforcing fisheries laws. Fishways were required for milldams in New Brunswick, yet none of the more than eight hundred milldams in the colony in 1851 possessed provision for fish passage. Illegal netting operations also remained widespread, eliciting a frustrated comment from Lieutenant Campbell Hardy, an officer of the Royal Artillery with a penchant for salmon fishing, who wrote in *Sporting Adventures in the New World:* "Netting in fresh water is forbidden. This practice, however, . . . is carried on to a frightful amount; the salmon being swept out of pools, and off their spawning grounds when in the very act of spawning. . . . For the greatest possible present gain, all future benefit is overlooked" (1855, vol. 2, 112). Hardy is also attributed with commenting on the general public's indifference to the plight of the salmon: "The spirit of wanton extermination is rife; and it has been well remarked, it really seems as though the man would be loudly applauded who was discovered to have killed the last salmon" (Dunfield, 1985, 135).

Though the importance of protecting the fishery was widely acknowledged, the process of doing so was ineffective in no small measure because it deferred to local interests. R. W. Dunfield, who in the 1980s compiled colonial and nineteenth-century accounts of salmon in Canada and New England, related the story of how Moses Perley, a lawyer commissioned in 1846 by the provincial government to investigate the state of New Brunswick's salmon, found illegal fishing rampant up and down the Nepisiguit River. Shocked by the obvious abuses, he requested that the local authorities take immediate action. The county Sessions Court at Bathurst, professing they had always protected the salmon fisheries, sent deputies upriver to investigate Perley's allegations. The posse did not return empty-handed. Though no poachers were apprehended, they managed to capture twelve dozen salmon.

The few fish wardens entrusted with enforcing fishing laws were notorious for looking the other way, their failure to carry out their responsibilities all but ensured by the lack of support from both the public and legislatures. In the few cases where miscreants were convicted, the courts mocked enforcement efforts with light penalties. Not that it really mattered. Enforcement was virtually impossible, given the realities of travel and communication. A single fish warden could not possibly patrol a territory covering hundreds to thousands of square miles. Many fish wardens didn't really know what they were supposed to do other than collect their annual stipend. Dunfield relates how in 1866, only two fisheries officers in the province of New Brunswick had copies of the 1851 Fisheries Act they were charged with enforcing.

As in England, the causes of declining salmon runs did not go unnoticed, or unpublicized. In 1864, a Vermont lawyer, George Perkins Marsh, published *Man and Nature*, a book that examined the effect of humans on the earth and its inhabitants. Marsh drew on his observation of changes in the landscape of his native New England and saw the destruction of the salmon as a tragic outcome not only of a proliferation of dams but of the impacts of the timber industry, pollution by mills and factories, and soil erosion that buried streams in fine sediment that clogged spawning beds.

Viewing man as a major force reshaping Earth's surface, Marsh advocated societal action to restore the declining fisheries:

> The restoration of the primitive abundance of salt and fresh water fish, is one of the greatest material benefits that, with our present physical resources, governments can hope to confer upon their subjects. The rivers, lakes, and seacoasts once restocked, and protected by law from exhaustion by taking fish at improper seasons, by destructive methods, and in extravagant quantities, would continue indefinitely to furnish a very large supply of most healthful food. (1864, 118)

Marsh recognized that salmon were adapted to their home streams and felt that salmon were particularly vulnerable to changes in their surroundings, stating,

> Fish are more affected than quadrupeds by slight and even imperceptible differences in their breeding places and feeding grounds. Every river, every brook, every lake stamps a special character upon its salmon, . . . which is at once recognized by those who deal in or consume them. . . .

[A]lmost all of the processes of agriculture, and of mechanical and chemical industry, are fatally destructive to aquatic animals within reach of their influence. (1864, 121–22)

Marsh cited a litany of harmful effects of human actions on salmon, including changes in the beds and currents of rivers, mechanical disturbance of spawning grounds, changes in stream temperature due to clearing of streamside forests, milldams, sawdust from lumber mills, and pollution from industrial operations. He viewed with alarm the unwitting destruction of natural systems as a result of the cumulative effects of small impacts taking place everywhere across the land. An instant success, Marsh's book became a cornerstone of the American conservation movement.

By the late 1860s Maine's salmon rivers were so depleted that the state appointed two commissioners of fisheries, Nathan Foster and Charles Atkins, and charged them with conducting a survey of the disaster. After two years touring salmon fisheries around the state they reported that "many rivers have become almost entirely depopulated, and we do not think one in the State can be found that has maintained its fisheries in a fair degree" (Commissioners of Fisheries 1869, 71–72).

All across Maine, Foster and Atkins found dams and barriers impassable to salmon, evidence of overfishing, and pollution of streams and rivers by sawmills and paper mills. They recounted story after story about how salmon had disappeared from rivers throughout the state after dams were built without regard for fish passage.

Foster and Atkins urged that state fisheries laws be revised and proposed a five-year ban on salmon fishing in Maine's four largest rivers, the Androscoggin, Kennebec, Penobscot, and St. Croix. Echoing Alexander Fraser's recommendations for saving British salmon three decades earlier, they concluded:

> To restore the sea fish [salmon] to our waters these conditions are essential:
> First, that fishways be built over all impassable dams.
> Second, that excessive fishing be prevented.

Third, that the waters be not poisoned.

Fourth, that in some cases fish be bred in the waters to be restocked.

(Commissioners of Fisheries 1869, 76)

Although the commissioners considered all four of their recommendations as essential for saving Maine's salmon, the state aggressively pursued only the fourth recommendation, which did not run afoul of powerful vested interests.

In addition to recognizing the multiple causes of salmon declines, Foster and Atkins saw firsthand how the complexity of the issue made it difficult to convince commercial interests to take actions to reduce impacts to salmon. They recounted how until 1825 the annual catch at Salmon Falls on the St. Croix River had been about 18,000 salmon, but that after construction of a dam without a fish ladder near the mouth of the river the annual catch for the whole river dropped to just several hundred fish. The obvious solution was to retrofit the dam with a fish ladder. However, the commissioners recognized that it would be

nearly useless to build fishways and attempt to restock the river with salmon unless some stringent law is enforced with reference to the time, manner and place of fishing. The drift nets are now thrown as near to the Union dam as a boat can venture, within a few feet of the falling water. . . . Every salmon that approaches the dam is doomed. If there were a fishway but few would ever succeed in entering it. (71)

Action was stymied because a fish ladder alone would not solve the problem. The fact that various interests responsible for depleting salmon runs could easily find pretexts to blame each other complicated efforts to restore depleted runs. Mill owners did not care to be singled out as the parties responsible for the demise of Maine's salmon. Foster and Atkins were "happy to state that the mill owners evinced a very laudable readiness to do their part by constructing and maintaining fishways, but they with reason objected to being compelled to incur the necessary expense, unless sufficient restriction be put on the fishing below to insure that the outlay shall not be in vain" (71).

Fishermen had their own ideas about who was to blame, preferring to see dam operators and competing fishermen as responsible for the decline of salmon. The commissioners commented:

> It seems to be a general opinion among the fishermen of the Kennebec that the time has come for a radical change in some direction or other to save the fisheries from destruction. There is, to be sure, very little harmony in their views of a remedy. The different classes of fishermen [i.e., those using other types of fishing gear] too generally exonerate themselves from all blame and throw it upon the other classes. (29)

The propensity for fishermen and dam owners to blame each other for decimating Maine's salmon helped keep the political process from addressing the causes of the decline of salmon runs. Instead of adopting Foster and Atkins's four-part program, or their proposal for a simple five-year moratorium on fishing, the state opted for measures more palatable to commercial interests. Hatcheries were to save the day.

Atlantic salmon conservation efforts began in the 1870s, at first concentrating on hatchery production using eggs from Maine's Penobscot River. Millions of salmon fry were released in the Connecticut River. As restocking efforts began to rebuild runs, fishing intensity rose in the unregulated fishery. In response, new legislation prohibited targeting salmon and required that any fish unintentionally caught be released immediately. Poachers, however, could claim that a fish had died before it could be returned to the water, at which point it would be wasteful not to have kept it. A steady stream of accidentally caught salmon fed markets for fresh salmon in New York City. Recognizing that fishing interests were undermining restoration efforts, the state legislature stopped funding restoration efforts, believing it would be a waste of money should fishermen take all the fish before any had spawned. In 1887, the U.S. Commission of Fish and Fisheries reported that Connecticut River salmon were "practically extinct."

A similar story played out on the Merrimack River. The U.S. Fish Commission began rehabilitating neglected fishways and through the 1880s salmon began to return to the river in increasing numbers. In

Salmon fishermen
in New England
in the 1880s.

addition to retrofitting dams with functional fishways, the river was stocked repeatedly with thousands of juvenile fish starting in 1867. Enough adult salmon began returning to support hatchery operations and the runs started to revive. High flows in 1890 prevented fishermen from setting salmon nets in the river, allowing the largest number of spawners in memory to ascend the river unmolested. But the Merrimack River salmon disappeared completely by the early twentieth century owing to the continued proliferation of dams, rampant water pollution, and aggressive fishing at the river's mouth.

Maine's salmon runs were in serious trouble by the end of the nineteenth century. An 1887 government report estimated that 90 percent of the productive capacity of Maine's rivers had been lost to blockages from dams. Salmon remained in less than half of the rivers that originally supported runs in Maine. By 1908, salmon were so rare in portions of the Penobscot River that newspapers reported when one was caught. In the end, hatcheries simply could not sustain runs without a supply of eggs from returning adult salmon.

Interest in river restoration and salmon recovery more or less languished in New England until the 1960s. Passage of the Anadromous Fish Conservation Act by Congress in 1965 promised new federal support to rebuild New England's salmon, renewing interest in salmon recovery throughout the region. Over the next few years state fish and game commissions and federal agencies began cooperative fishery restoration programs. Hundreds of millions of dollars were invested in rebuilding Atlantic salmon stocks and reestablishing runs in rivers where they had been extirpated. The sale of wild Atlantic salmon was banned to remove the incentive for commercial fishing while spawning runs were rebuilt.

Sadly, Atlantic salmon-recovery efforts stalled out after a few decades. New England's wild salmon totaled less than six thousand fish in the early 1990s and about a thousand fish by 2000. Suspicion focused on ocean fishing as the culprit that was decimating salmon stocks in spite of efforts to improve river conditions.

The Labrador Sea is the principle ocean feeding grounds of North American Atlantic salmon. The primary area extends from the east coast of Labrador, north of the Grand Banks and Newfoundland to the west coast of Greenland. Ocean circulation patterns make the Labrador Sea an immense mixing zone between the cold Labrador Current and warmer North Atlantic waters. This great oceanic blender supports a highly productive marine fauna. Here Atlantic salmon from Europe and North America mingle to gorge on capelin (a 6-inch long barracuda-like fish), sand eels, herring, squid, krill, and lots and lots of amphipods, shrimplike crustaceans that give salmon flesh its reddish color. Most Atlantic salmon spend one to two years feeding in these waters before returning home to spawn. Because of the convergence of salmon from rivers in Europe and eastern North America, it is impossible to know where a salmon caught in the Labrador Sea originated.

Greenland, the eastern boundary to the Labrador Sea, is a semiautonomous state under Danish administration. Not knowing where salmon that they caught in the waters off Greenland came from didn't bother the Danes. After all, few if any were coming from Denmark. By the early 1960s the Greenland salmon fishery had expanded enough to attract the attention of fishing interests with huge oceangoing trawlers.

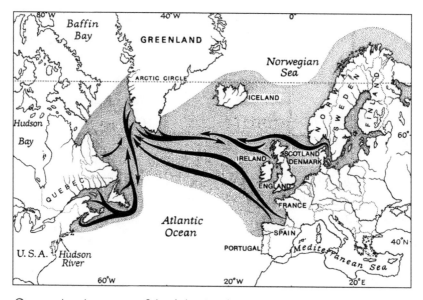

Ocean migration routes of the Atlantic salmon.

Introduction of nylon nets that were invisible to fish made ocean fishing quite efficient by the 1960s. Though illegal in most countries, such nets were favored by the Danes. Most important for the development of a high-seas salmon fishery, advances in freezing technology made it possible to get salmon from Greenland to overseas markets. Between 1960 and 1967 the annual catch off of Greenland grew from under 20,000 to almost 500,000 salmon.

Explosive growth of the Greenland salmon fishery alarmed conservation and fishing groups alike. An indiscriminate open-ocean fishery could wipe out depleted U.S. salmon stocks and further stress vulnerable runs in Britain and Canada. In 1966, T. B. Fraser, the president of the Atlantic Salmon Association, a Canadian salmon conservation organization, called for a ban on high-seas salmon fishing. Other organizations soon joined the chorus for restraining ocean fishing. The international body authorized to propose such changes was the seventeen-member International Commission for Northwest Atlantic Fisheries (ICNAF). Only five of the member states produced salmon in their rivers, which made it difficult to rally support for restricting the ocean fishery to protect on-shore investments in salmon conservation.

At a 1968 ICNAF meeting, the Canadian delegation proposed freezing the growth of the west Greenland salmon fishery. Citing the increased danger of extinction for Atlantic salmon presented by the rapid growth in the ocean fishery, the British and U.S. delegations supported the Canadian proposal. In an unusual Cold War–era alliance, the Soviets concurred that protection of salmon at sea was critical for salmon conservation. But the Danes, Icelanders, and Norwegians blocked the resolution, claiming that there was not enough scientific evidence to connect ocean fishing with the condition of particular depleted or threatened salmon runs. They also knew that such a connection would be very hard, if not impossible, to demonstrate conclusively for quite some time.

A year later, a resolution calling for a complete ban on ocean fishing for Atlantic salmon was passed under considerable U.S. pressure, and over the objections of Denmark and West Germany. The official U.S. position, delivered at the June 1969 ICNAF meeting, provides a clear statement of the basis for opposition to open ocean salmon fishing.

> The productivity of salmon runs from different rivers varies, and individual runs fluctuate from year to year, usually independent of each other. Coastal fisheries can be managed to allow adequate escapement for all runs, but this type of conservation can not be practiced on the high seas. . . . Immediate action is needed which will ensure coastal states that their careful efforts to conserve limited salmon resources will not be defeated by lack of conservation in ocean waters. (Buck, 1993, 59–60)

Denmark simply ignored the resolution and doubled its catch.

Indignation grew in Britain and her former colonies over Danish intransigence. The Danish government brushed off appeals from the British prime minister, Harold Wilson, and salmon conservation organizations. Concerned over the impact of ocean harvesting on recovery of New England salmon, in 1970 a group of well-connected New Englanders formed the Committee on the Atlantic Salmon Emergency (CASE) to restore Atlantic salmon in U.S. waters and advocate a complete ban on open-ocean salmon fishing. Richard Buck served as CASE's chairman in this opening shot in a years-long Salmon War.

CASE orchestrated a publicity campaign targeted at informing the American public of the Danish threat to the Atlantic salmon. The Danes continued to profess support for conservation in general but maintained that data were insufficient to demonstrate that their high-seas fishing threatened salmon runs. CASE countered that it would be many years before definitive data could be collected that would provide proof of their position, yet it seemed absurd to delay action until scientists were 100 percent certain of something as obvious as the fact that a fish taken at sea is one less that makes it back to the river. As the Danish high-seas catch increased, the Canadian catch of Atlantic salmon dropped sharply, falling almost 50 percent between 1969 and 1971. Refusing to acknowledge any connection between their big hauls and the alarming Canadian declines, the Danes kept on fishing.

In 1971, CASE enlisted the help of celebrities to publicize their efforts to convince the Danes to restrain their salmon fleet. One was Bing Crosby, who was part Danish. An enraged Danish government banned the sale of Crosby's records, only to discover that he was incredibly popular in Denmark. The move backfired and produced an outpouring of public support for both the crooner and the salmon. Moves to boycott Danish goods gathered momentum in the United States.

In February 1971, Thomas Pelly, a congressman from Washington State, made a move designed to ride the wave of public sentiment. He introduced a bill in the House of Representatives to protect salmon of North American origin. The time was right for such a move. Congressmen from New England introduced two similar bills. Richard Buck, the chairman of CASE, eloquently testified before the House Subcommittee on Fisheries and Wildlife Conservation and argued for banning open-ocean fishing for salmon.

In the high seas feeding areas, salmon stocks are inextricably intermingled. They come from different spawning streams, different river systems, different nations, different hemispheres. No man, and no type of fishing gear yet known to man, no method of control, can separate them out. Thus high seas fishing takes indiscriminately from perhaps the very river runs needing particular protection, and results in absolutely no rational or scientific means of conserving basic stocks or ensuring adequate escapements for spawning. . . .

Danish fisheries authorities and the Danish government profess great interests in conservation measures, and it is assumed therefore that they certainly understand this basic tenet of proper management of salmon stocks. Yet they never acknowledge this basic principle, probably because to do so would place them in an indefensible position. (*Congressional Record* 1971)

Influenced by such testimony, the language in the three bills was changed to give the secretary of commerce broad authority to impose an embargo on products from countries conducting fishing operations that compromise the effectiveness of international fishery conservation programs. The final bill sailed through both the House and Senate and was signed into law by President Nixon in December 1971.

The Danes recognized that the new law authorizing trade embargoes was aimed at them. Under pressure from exporters who feared losing U.S. markets, whether to boycotts or embargoes, the Danish government quickly sought negotiations. Secretary of State Henry Kissinger, concerned over possible effects on the solidarity of the NATO alliance, demanded to be briefed twice a day on the salmon negotiations. In February 1972 the Danes agreed to phase out the Greenland open ocean fishery, with a complete ban beginning in 1976. Although the informal agreement could be abrogated at any time, it finally appeared that the ICNAF resolution would be implemented.

The agreement had a ripple effect. Two months later, Norway announced it would phase out and then ban high-seas salmon fishing off its coast. Several weeks after that the Canadians announced support for a complete ban on ocean fishing for salmon. Then Canada went even further and banned commercial fishing for salmon in New Brunswick and Nova Scotia, including drift netting in Canadian waters. The ban affected just under half of Canada's commercial Atlantic salmon fishery. Originally intended to last a single life cycle of the salmon, the ban remains in effect. At their next meeting ICNAF adopted the provisions of the U.S-Danish agreement. In December 1972 Denmark passed a law implementing the agreement. The Salmon War appeared over. Victorious, CASE disbanded.

But the Danes did not give up easily. As open-ocean fishing was phased out, near-shore fishing in Greenland, technically in Danish waters, increased. In the first two years after the U.S.-Danish accord, which contained a provision allowing but limiting the near-shore catch, the Danes' Greenland catch exceeded the negotiated limit by 43 percent. At the next ICNAF meeting the Danes requested that their near-shore Greenland limit be raised by a third. They maintained that the request was warranted because the collapse of the cod fishery because of overfishing presented an economic hardship on native Greenlanders. According to the Danes' logic the economic dislocation resulting from destruction of one fishery justified overexploiting another. The ICNAF had been based on seeking solutions by consensus, but the organization, apparently ineffective at conserving the fish it was intended to protect, disbanded by the end of 1976.

Concerned over the failure of international salmon management, Senator Warren G. Magnuson of Washington and Representative Gerry Studds of Massachusetts sponsored the 1976 Fishery Conservation and Management Act, which would create a 200-mile-wide enforceable fishery conservation zone off the two coasts of the United States. Signed into law by President Gerald Ford, the act empowered the United States to kick foreign fishing fleets out of the coastal waters and off the continental shelf of North America. It also asserted U.S. dominion over "all anadromous species that spawn in U.S. water, throughout their migratory range beyond the zone, except when they are in another nation's territorial sea." In effect the United States asserted a right to all salmon that spawned in U.S. waters unless they wandered into another country's direct jurisdiction.

The huge early 1970s harvests by the Danes in Greenland that peaked in 1972, when 750,000 salmon were taken, had depleted stocks enough to sharply reduce reproduction. Now a population "echo" was noticeable: The return of low numbers of salmon to their river spawning grounds, due to overfishing at sea, was resulting some years later in falling ocean catches back at Greenland. In 1979, Canada's salmon catch plunged by more than half from the already depressed 1978 level. Concern grew over the potential for a negative feedback loop in which sustained overfishing depressed the number of spawners, leading to a cycle that ratcheted down already depressed stocks. Mortality was racing well ahead of reproduction.

A formal international ban on ocean fishing for salmon was proposed at the 1978 Atlantic Salmon Symposium in Edinburgh. Five years later, after much political wrangling, the North Atlantic states—including Denmark—signed a convention that committed them to conserving, restoring, and enhancing salmon in the North Atlantic. The following year, in 1984, the North Atlantic Salmon Conservation Organization (NASCO) convened its first meeting in Edinburgh. The purpose of the organization was to set quotas for the ocean salmon harvest.

In practice, the catch fell faster than the quotas. The world harvest of Atlantic salmon fell from 2.9 million fish in 1967 to 1.7 million fish in 1982, a drop of over 40 percent in only fifteen years. Then it fell even faster. Spawning runs in Canada dropped precipitously in 1983, and the Canadian catch had dropped by more than a third since 1980. Greenlanders were only able to catch a quarter of their allotted quota in 1983. Between 1986 and 1992 the worldwide catch of salmon dropped again by half. Salmon populations were crashing.

Initially the harvest targets set by NASCO were based on limited information and there was substantial resistance to cutting back on the catch. By 1992 scientists came up with a method of forecasting salmon abundance based on sea-surface temperatures, the number of juvenile salmon going to sea, and the size of the feeding area in question. The next year, NASCO agreed to base harvest allocations on scientific advice. Population models and research led scientists to advise a complete closure of the Greenland salmon fishery in 1995. This was unacceptable to the Danes, who expressed their intention to vote for a 77-ton quota. As NASCO operated by consensus, a single dissenting vote would have resulted in no quota at all, leaving the fishery an unregulated free-for-all. After much argument NASCO adopted the Danish position, ignoring the scientific advice it had pledged to follow.

Danish fishing interests also figured out other ways to circumvent international conventions on salmon fishing. In the winter of 1989, Danish ships registered in Poland and Panama caught an estimated 630 tons of salmon in international waters. These ships registered under foreign flags delivered their salmon catch to Poland, whence the fish were shipped to Switzerland. None of these salmon counted toward Denmark's quota under NASCO.

Canada drastically cut back both commercial and sport fishing in 1984. The United States went a step further and banned the sale of Atlantic salmon caught in its waters, allowing angling only for private consumption. But the British continued to allow intensive fishing in their coastal waters. More than 90 percent of the salmon caught by British nets off northeast England were salmon heading home to spawn in Scottish rivers. John Gummer, the British minister for agriculture, food and fisheries, defended coastal fishing as a traditional fishery, ignoring concerns that its catch had grown thirtyfold in as many years. Richard Buck proposed phasing out the British driftnet fishery over a five-year period, as had been done in Greenland. Although the British opposed the indiscriminate open-ocean fisheries in the North Atlantic, they wanted to protect their domestic coastal fishery by allowing salmon to be taken in near-shore waters.

Looking back over several decades of involvement in Atlantic salmon–recovery efforts in the early 1990s, Richard Buck saw that rational salmon management was not simply a technical problem but was far more a political one. He ascribed the difficulty in getting countries to conservatively manage salmon stocks to the simple fact that fisheries ministers could not afford to alienate commercial fishermen. Ironically, Buck notes that sustained American pressure to curtail overfishing was possible because action was not prevented by the political power of commercial fishing interests. American commercial fishing for Atlantic salmon was already history because there were hardly any fish returning to American waters.

There are real economic incentives to protect and restore Atlantic salmon to North American rivers. A 1990 Canadian government survey indicated that anglers spent over $240 million and created 49,000 seasonal jobs in the Atlantic Provinces. Restoration of wild salmon runs could provide a huge stimulus to a region hard hit by the economic fall-out of poor fishery management in the past. The stakes are even higher in efforts to recover North Atlantic salmon fisheries. Potentially worth billions for the fish alone, the value is far greater if one includes the economic stimulus of sport fishing in coastal and up-river areas. It is priceless if you include the value of people knowing that future generations will be able to fish in rivers full of salmon.

After the Salmon War of the early 1970s, the founders of CASE regrouped. Working to restore Atlantic salmon stocks in New England, they focused on construction of fishways around the dams that had destroyed salmon runs of the Connecticut and Merrimack Rivers. Once again, connections with Nixon administration officials proved decisive, and the Department of the Interior required dam operators to finally provide for fish passage. Meetings with top officials also overcame the main obstacle to restoration along the Connecticut River. The Army Corps of Engineers dropped its long-held position that water pollution would preclude successfully reestablishing anadromous fish so it was pointless to install fishways. Now, restoration was no longer officially labeled futile.

There are some success stories in New England and Canada where restoration efforts have concentrated on restocking depleted rivers. The Nepisiguit River is a dramatic example of the potential to restock even decimated rivers. Atlantic salmon were exterminated from the river in 1969 when heavy acid rain fell on tailings from an open pit mine. Toxic runoff enriched in lead and zinc killed all the juvenile fish in the river. Adult fish returning from the ocean avoided the river for five years, stretching the disaster across a complete life cycle and virtually eliminating salmon from the river. Eventually efforts to restock the river with fry and smolts from the Restigouche and Miramichi rivers reestablished runs of a few thousand fish by the late 1980s. Although habitat is plentiful and now in good condition, these runs have stayed small owing to poor marine survival thought to be caused by ocean fishing.

In December 1997 the National Marine Fisheries Service and the Fish and Wildlife Service withdrew a proposed listing under the Endangered Species Act the last naturally reproducing runs of Atlantic salmon in the eastern United States. Bowing to pressure from the state of Maine, the federal agencies accepted the state-proposed Atlantic Salmon Conservation Plan to protect runs extending from the Kennebec River to the St. Croix. Two years later, in October 1999, the agencies released a status review that confirmed that these last runs of wild Atlantic salmon continued to decline under that plan. Survival rates for both juvenile and adult salmon were lower than expected and more stringent measures were needed to reverse the downward trend. In November 2000 Atlantic salmon in eight rivers in Maine were listed under the Endangered Species Act.

The listing resulted in a political uproar over the status of Maine's remaining Atlantic salmon. Local interests, fearful of the economic impact of an ESA listing, argued that the remaining populations of Atlantic salmon were derived from previous hatchery stock, and from fish farm escapees—in other words, their logic went, only wild fish were protected under the ESA.

A National Academy of Sciences committee was convened to address the issue, and it found compelling evidence that Maine's remaining salmon are derived from remnants of wild stocks. Despite the extensive introductions of foreign hatchery fish to Maine's rivers, populations in different rivers were genetically far more distinct from one another than would be expected if the hatchery programs had been as successful as fisheries managers believed. Surprisingly, the genetic composition of Maine's wild Atlantic salmon indicated little long-term genetic impact of intensive hatchery programs. The NRC committee concluded:

> [T]he hatchery fish have not displaced the local gene pool because of the poor success of historical hatchery stocking and the likelihood that Canadian fish were poorly adapted to Maine streams.
>
> [D]espite 130 years of stocking . . . and releasing about 120 million Atlantic salmon, the systematic decline in run sizes has not been reversed. That raises the question of whether hatchery stocking has ever had a substantial impact on populations of Atlantic salmon in Maine. (National Research Council 2000, 19)

In the end, New England's reliance on hatchery programs to rebuild Atlantic salmon stocks produced few tangible results.

Like the saga of the British salmon, the history of the Atlantic salmon in New England has relevance to our modern salmon crisis in the Pacific Northwest. Perhaps the most striking lesson is that hatcheries can only be effective to sustain a fishery if habitat also remains in good shape. Local control of both fishing and salmon passage regulations proved no more effective in New England than in England. Poor enforcement of well-construed laws and regulations undercut salmon conservation efforts. And finally, the political and regulatory process tends to defer to powerful industries at the expense of public salmon fisheries.

The primary obstacles to reversing the Atlantic salmon's long slide toward extinction always have been, and still are, sociological and political. The techniques and experience of New England's fishermen transferred well to the Pacific Northwest. Before methods for preserving and shipping salmon were developed, the salmon that were the lifeblood of the native subsistence economy in the Pacific Northwest had no commercial value to the new arrivals. But once canneries turned salmon into a commodity, the understanding of what drove salmon declines in the East was put to little use in managing the fishery. Just like Britain, New England traded its salmon for milldams and factories. The Pacific Northwest was even quicker to cash in its bounty of salmon in a water-borne gold rush.

WESTERN SALMON RUSH

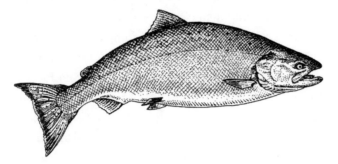

If you use your fish wisely you can have them forever.

Miller Freeman, Editor, *Pacific Fisherman*, 1952

ONE THING THAT XENA THE DOG AND I AGREE ON IS THAT salmon are great to eat. For my money, there is only one way to eat them—cooked on an open flame. Xena has a more open mind and even enjoys rotting carcasses when she gets the chance (and can get away with it). The fresh salmon BBQ is a Pacific Northwest tradition that partially compensates for living in a land of endless rain. So I never thought about buying canned salmon, until one day when I found myself starring at tins of sockeye stacked floor to ceiling in a duty-free supermarket on a converted U.S. Air Base in the Philippines. Here I was in the tropics, studying rivers too hot for salmon and yet able to buy tins of them for my lunch.

A few weeks later, at the end of a research trip down the Nisqually River, my colleague Brian Collins and I pulled our canoe out just upstream of at least twenty fishermen lined up along the banks. Shoulder-to-shoulder these plaid-bellied sentinels blockaded the mouth of the river. This was getting to be a familiar scene, a rerun of the one at Kennedy Creek. So the next time I ran into Phil Peterson I asked him about how common it was to have an honor guard of fishing poles welcome salmon back to their home rivers. He astounded me not by telling me that, yeah, it was common (I was clueing into that already) but by telling me that none of this sport catch is counted in official tallies of fishing pressure on salmon stocks. I recall thinking that it's no wonder salmon runs are crashing. Commercial fishing and open-ocean harvests catch the maximum allowed, and then a gauntlet of rod fishermen take more of the remaining fish at the river mouths. From the 1970s to the 1990s, fishermen took 60 to 90 percent of the runs each year.

The wall of fishermen guarding the entrance to the Nisqually River reminded me of trawlers I saw off of Chichigof Island in Alaska some years ago when I was there to study small streams in a research area in an old-growth forest. Access to this remote part of Southeast Alaska was by floatplane or boat. On our way back from a shower-and-beer-run to the nearest town, we saw fishing boats virtually, although perhaps not technically, blocking stream inlets with their nets. They parked, spread a net just beyond the mouth of the stream, and took anything that tried to run past. It seemed miraculous that any salmon would make it back upstream to spawn. My compatriots advised me that the fishing season only opens for limited periods, providing large numbers of salmon opportunities to scoot past the nets.

Now with recent advances in commercial sonar fishermen don't need to wait for salmon to come to them, for they can find schools of salmon out at sea and round them up without blockading the streams. They don't even see the fish until they pull up the nets. That's a lot of progress since Georg Wilhelm Steller, a German-born adjunct lecturer in natural history at the St. Petersburg Academy of Sciences, first described the different species of Pacific salmon.

Steller was ordered in 1737 to join Vitus Bering's second expedition in the Russian Far East. After spending nine years exploring Siberia, Steller became the first scientist to set foot in Alaska. On his way to

Alaska Steller observed hordes of salmon in Siberian rivers and described the five major species of Pacific salmon, calling them by their local Russian names. He noted that as fingerlings they went to sea for one or more years and grew to their full size before returning to spawn close to where they were born. He also described how huge runs of salmon would displace enough water to cause streams to overflow their banks. Sadly, Steller died on the way back to St. Petersburg. Another member of the expedition, Stepan Krashnennikov, published the first study of the Pacific salmon in 1755, based in part on Steller's journals.

Early explorers with less scientific curiosity focused on the native fishery and economic potential represented by the hordes of salmon in rivers of the Pacific Northwest. On August 3, 1805, Meriwether Lewis became convinced that the Corps of Discovery had crossed the continental divide and reached land that drained to the Pacific Ocean when an Indian gave him a piece of fresh salmon. On their way down the Columbia River the explorers passed extensive native fishing operations. Impressed by the vast quantities of split salmon drying on scaffolds outside of huts and lodges William Clark noted in his diary that "the multitude of this fish is almost inconceivable." Less than a decade after Lewis and Clark reached the mouth of the Columbia River, members of the Pacific Fur Company reported on the commercial potential of the salmon fishery, noting that native fisherman could easily furnish 1,000 tons a year. Their report trumpeted that a single man could catch five hundred fish in a day's work at the peak of the run. Word of the remarkable western salmon runs quickly spread back east where salmon runs were dwindling rapidly.

The vast runs of Pacific salmon came into the public eye just as commercial fishing on the depleted eastern runs began to expire. Like the salmon of eastern North America that had drawn the attention of Scottish and English fishermen, the western salmon now proved enticing to easterners. Salmon exports from the Pacific Northwest began in 1829, when a brig from Boston arrived at the Columbia River and sailed for home with fifty barrels stuffed full of huge fish. Exports of salted Puget Sound salmon soon followed.

You couldn't have found a better spokesman for the commercial potential of the region's fisheries than Charles Wilkes, commander of the

United States Exploring Expedition to the Pacific in 1838–42. Wilkes described Indian fishing at The Dalles, Oregon, in his influential and widely reprinted official report: "The men are engaged in fishing and do nothing else. . . . [I]t is not uncommon for them to take twenty to twenty-five salmon in an hour" (Wilkes 1844, vol. 4, 411). Wilkes also recorded how at Kettle Falls the Indians caught salmon in a large basket they lowered into the river: "This basket, during the fishing season, is raised three times in the day, and at each haul, not infrequently contains three hundred fine fish" (472). Wilkes's portrayal of a region with an endless supply of the finest salmon also emphasized that the superabundance of fish made them cheap; the going rate for a large salmon was ten cents.

Other early observers recorded similar scenes. James Swan, a pioneer and one of the Pacific Northwest's earliest ethnographers, described native fishing near the mouth of the Columbia River in 1853. Using a net made of spruce-root fibers and outfitted with cedar floats and notched beach pebbles for weights, members of the Chinook tribe employed a three person technique. Two men in a canoe with one end of the net drifted downstream while their partner walked along the bank with the other end. Sometimes nothing was caught, but "frequently a hundred fine fish of various sizes are taken at one cast of the seine" (Swan 1857, 107).

As in colonial New England, the supply of salmon in the Pacific Northwest appeared inexhaustible. Recognized as an aquatic gold mine wanting only access to a market, the varieties of Pacific salmon nonetheless presented an enigma to European arrivals to the region. Confusion dominated debates over how many species there were and their relationship to the Atlantic salmon. At first, the Pacific salmon were considered varieties of the genus *Salmo*, but recognition of key differences led to a long and confusing debate over salmon taxonomy.

When the German naturalist Johann Julius Walbaum updated the *Encyclopedia of Fishes* in 1792 to include the newly described Pacific salmon, after the genus name *Salmo* he replaced the species name *salar* with Steller's Russian species names: *tshawytscha* (chinook), *kisutch* (coho), *nerka* (sockeye), *keta* (chum), and *gorbuscha* (pink). Half a century later, Washington Territory's governor Isaac Stevens's report on the potential route for a northern railroad included a section, written by the American ichthyologist George Suckley, on the natural history of salmon that confessed befuddlement over both the number of species of salmon and their distinguishing characteristics. In 1861, Suckley distinguished the Pacific salmon from the Atlantic salmon as a separate subgenus of *Salmo*. On the basis of the appearance of breeding males he proposed the name *Oncorhynchus*, "snout-nosed," for the Pacific salmon. Within a decade some considered the Pacific salmon a distinct genus, but confusion about relations between the Atlantic and Pacific salmon and their relation to trout continued for almost another century. Paleontologists Ralph Stearley and Gerald Smith finally settled the ongoing taxonomic argument in 1993, more than two hundred years after Bering had first named them.

As a young man, Ezra Meeker, a pioneer and early promoter of Washington Territory, saw firsthand the abundance of salmon in the Columbia River in the territory's early days. In 1870 Meeker described how "Two men with their boat and net will average their twelve hundred pounds in one night. . . . [I]t is thought impossible to cut short the supply, as millions [of salmon] pass up the river. . . . In the first run are found some fine specimens weighing seventy pounds" (21).

In some ways the abundance of salmon in the region was simply a nuisance to European immigrants. In a practice reminiscent of colonial New England and medieval England, settlers found a use for the surplus of salmon. Too common to have commercial value as food, salmon did provide free, high-quality fertilizer. Farmers pulled wagonloads of salmon from streams during spawning season and used them as fertilizer or hog feed. Reminiscing as an old man in the early 1920s, Ezra Meeker recalled the abundance of salmon on the Puyallup River, Washington, seventy years before, in the 1850s: "I have seen the salmon so numerous on the shoal water of the channel as to literally touch each other. It was utterly impossible to wade across without touching the fish. At certain seasons I have sent my team, accompanied by two men with pitchforks, to load up from the riffle for fertilizing the hop fields" (1921, 280).

Salmon became a valuable commodity only after the arrival of cannery technology allowed preserving and shipping them to distant markets. Like many technological advances, the art of canning developed from military applications. During the French Revolution, Napoleon Bonaparte offered a 12,000-franc prize to anyone who devised a way to preserve food for sailors on long voyages. Nicholas Appert claimed the prize in 1809 by developing a method for sealing cooked food in glass jars. Sturdy tin cans quickly replaced the fragile glass containers and enabled shipment of salmon to markets around the world. Commercial salmon canning began in Scotland in the 1830s and spread to Maine by the early 1840s. The technology was just maturing as the attention of merchants turned to the Pacific salmon, and it allowed frontier entrepreneurs to reap huge profits. Once this occurred, competition to cash in fueled the first phase of the dramatic collapse of the Pacific salmon. Rampant overfishing led to sequential depletion of major stocks in the western United States, beginning in California rivers and proceeding north through Oregon and the Columbia River and then on to Puget Sound. The timing and details of the story for these runs differ, but they share a similar outline.

During the Gold Rush, approximately 1848 to 1864, salmon fed miners swarming along the banks of California's rivers and streams searching for gold. The miners mostly ate fresh or salted fish, and entrepreneurs like John Sutter, at whose mill James Marshall's cry of "Eureka!" started the Gold Rush, ran salmon fishing operations to feed the swelling population. However, the 49ers' fishery paled in light of what followed the advent of canneries and the access they allowed to larger markets.

Seeing the lucrative potential for harvesting California's salmon, Andrew Hapgood and two of the three Hume brothers, George and William, established the first cannery on the Pacific Coast in the town of Washington on the banks of the Sacramento River in 1864. In its first year of operation, the new cannery produced two thousand cases of salmon, each case containing forty-eight one-pound tins. The operation was plagued by faulty equipment resulting in the loss of half of the first year's pack (the term for the amount of salmon a cannery managed to stuff into cans). Hapgood, Hume & Co.'s bad luck did not stop there. After two more years of substantial effort and little profit, the Hume brothers abandoned California to establish the first cannery on the Columbia River. Their business doubled in the first year. Dozens of other canneries quickly followed. Within ten years salmon canning on the Columbia River grew from 4,000 to 450,000 cases a year, and to 630,000 cases by 1883. By then the initial astronomical profits had been brought down to earth by increased competition for a limited supply of fish. The Hume brothers were already millionaires.

Thirty years later, R. D. Hume, the third and youngest Hume brother, published *Salmon of the Pacific Coast*, a somewhat repentant pamphlet that detailed the growth of the canning industry, the overexploitation of salmon, and the destructive effects of hydraulic mining, the timber industry, and dams on salmon runs of Oregon and California. Robert D. Hume was born in Augusta, Maine, in 1845 and witnessed the decline of the Atlantic salmon in the Kennebec River. At the age of eighteen he followed his elder brothers to Sacramento, and, later, once again when they went north to establish their business on the Columbia River. After the death of his wife, the youngest Hume relocated to the mouth of the Rogue River on the southern Oregon coast and set up a cannery that he ran from 1876 until he died in 1908.

Hume was keenly, perhaps painfully, aware of the overexploitation of the Pacific salmon, and he saw firsthand the destructive effects of changes in the landscape and rivers on the salmon. Concerned over ensuring a steady supply of fish for his cannery, he built the first salmon hatchery in Oregon. His concern extended to the future of the industry, as revealed by his 1893 pamphlet that, he said, he had written "[t]o call the attention of both producer and consumer to the danger of the total extinction of this most valuable of food fishes, and provide a simple method for their preservation" (Hume 1893, 5). Hume's pamphlet reviewed the dramatic changes that occurred across the region between 1867 and 1892, noting how at the earlier date their first cannery had difficulty finding employees, whereas at the later date every river along the Oregon and California coast had "one or more canneries located on its banks."

Hume tells of how though raised in a Maine fishing family he never tasted salmon until he left for the Pacific Coast. Salmon was "a luxury of which none but the wealthy could partake" in his native town of Augusta, Maine, and he thought it "doubtful if but few in that State of his age had ever seen one." Hume saw canned Pacific salmon as a blessing that could supply cheap nutrition to the poor.

Within a few years, Hume's vision seemed prophetic. The growing West Coast salmon canneries aggressively marketed their product overseas. Soon, salmon once again fed not only the British public, but outposts of the British Empire around the world.

Hume remained puzzled by the wanton destruction of both salmon runs themselves and the rivers and streams that sustained them. He recoiled from the collective strangling of the region's golden goose. Hume recalled the Kennebec River he knew in his youth, reflecting upon how it had been a "fine salmon stream" in colonial times and how "the mind is easily satisfied as to the causes which led to the almost total extinction of these fish."

He recounted how a dam at Augusta had been built without a fish ladder, blocking salmon from access to areas upstream, and how sewage, industrial wastes, and sawdust from sawmills polluted the river. Spawning grounds in tributaries were sluiced out by the transport of logs from intensive timber harvesting in their headwaters. And the lower part of the river was "lined with traps to such an extent as to render the escape of a fish almost an impossibility."

Hume contended that it was well known how these effects and others conspired to drive the salmon runs of the Kennebec River into commercial extinction—the Sacramento River had been full of salmon before sediment washed down from hydraulic mining destroyed the runs supplying Hapgood, Hume & Co.'s first cannery:

> The Sacramento river, prior to the introduction of hydraulic mining in 1853 was, during the running season, so plentifully stocked with salmon that no use could be made of but a moiety of the supply, and we have an illustration of the destructive force of this new agent [hydraulic mining] when we consider the fact that eleven years after its introduction the Sacramento river was practically rendered useless for commercial purposes as a salmon stream. (Hume 1893, 19)

The Sacramento was just one of many rivers of the Sierra Nevada where the Gold Rush destroyed salmon runs. Hydraulic mining in the Sierra Nevada dumped huge quantities of sediment into rivers draining the range. This mining technique involved pumping river water into a powerful hydraulic cannon that shot a jet of water to carve into ancient river gravel containing gold nuggets. The sediment blasted free from these deposits passed downstream; the small amount of gold remained with the miners. River after river filled in with the vast amount of sand and gravel blasted free by miners' hoses. By the late 1870s some streams were buried under more than 100 feet of debris.

The mining waste moved slowly down the rivers from the mountains to California's Central Valley. Even today, many Sierran streams are still cleaning themselves out, gradually sending the excess sediment on downstream. As the peak of the sediment pulse passed through Sacramento, the bed of the river rose enough to flood the State Capitol. By the time the wave of sediment crested, the riverbed rose above adjacent farmlands. Fearing their own ruin, farmers pushed through legislation banning hydraulic mining. By then it was too late to save Sierran salmon.

Northern California also had lots of salmon outside of the Sierra Nevada rivers. Eight years after Hapgood, Hume & Co. bailed out of the Sacramento River to chase Columbia River salmon runs, salmon canneries returned to the river north of the area devastated by hydraulic mining. By the early 1880s as many as twenty-one canneries operating on the Sacramento River packed as many as 200,000 cases of Sacramento River salmon. On the Northern California coast, salmon canneries operated sporadically from 1878 to 1904, but never packed more than 20,000 cases. Northern California salmon canneries resumed operations in 1909, rose to a peak during the First World War, and then declined until salmon canning ceased in 1933. In its biennial report for 1934–36 the California Division of Fish and Game predicted the eventual demise of California's salmon.

California's salmon fishery, the oldest and once the most important fishery in the state, continues to decline and it is inevitable that our two species of salmon, the king [chinook] and silver [coho], will become commercially extinct in California unless more protection is given them. ... Facts relating to the depletion of this once great salmon run as well as what should be done about it are well known. . . .

[T]oo large a portion of the run is caught before the salmon can get in the river [to spawn]. (1937, 47–49)

Even without the effects of hydraulic mining, overfishing pushed Northern California's salmon down the road toward commercial extinction. Almost half a century earlier, Hume's 1893 pamphlet had identified overfishing as the primary culprit responsible for depletion of the Columbia River fishery. He observed that the cannery output on the Columbia River dropped by "more than one-third" between 1883 and 1892. The decline was particularly acute in the prime adult chinook salmon, which had decreased by "more than three quarters." Hume questioned why the decline continued when the number of canneries was decreasing and a hatchery was turning out "millions of young salmon every year." Hume's fears for the long-term stability of the Columbia River salmon soon materialized as some runs began to disappear altogether.

Hume called for the state to prevent dams and traps from blocking salmon from their spawning grounds, to create a fish warden for the state with deputies in every county, and for canneries to establish hatcheries to sustain the harvest. Although Hume proposed to equip every cannery with a hatchery, he recognized that salmon stocks in different rivers were distinct and adapted to their home stream. Consequently, he advocated using stocks from streams similar to those into which they were to be released as the key to successful salmon breeding.

Less than two decades after Hume set up his cannery on the Columbia, Livingston Stone, the U.S. Fisheries Commission agent in the Pacific Northwest, explored the river looking for a place to establish a salmon hatchery. He complained of having a hard time finding a stream in which to collect enough eggs to support large-scale salmon breeding.

Twenty years ago, before the business of canning salmon on the Columbia was inaugurated, salmon literally swarmed up all the small creeks and little tributaries of the main river . . . but that day has gone by, probably forever. The vast number of nets that are being continually dragged through the water at the canneries on the main river during the fishing season catch millions of full-grown salmon on their way up the river to spawn, and of course reduce to a corresponding extent the number of parent fish that reach the spawning-grounds. . . . It is accordingly useless to look now to small streams which are subject to ordinary conditions for a large supply of salmon eggs, however abundant the salmon used to be in them in the former and better days of these salmon rivers. (Stone 1885, 242–43)

Mining of chinook salmon on the Columbia peaked in 1883 with a catch of almost 43 million pounds. As chinook landings began to decline, the catch of coho and sockeye rose to make up the difference. Anticipating the decline led the historian Hubert Bancroft to comment on the future of the salmon fisheries in his *History of Oregon*: "Nature does not provide against such greed, and it is doubtful if art can do it. The Government, either state or general, should assume control of this industry by licensing a certain number of canneries, of given capacity, for a limited period, and improving the hatcheries. Otherwise there is a prospect that the salmon, like the buffalo, may become extinct" (1888, 758). Even before cannery operations began to run out of fish, it was clear that commercial fishing for Pacific salmon was in jeopardy unless salmon received greater protection.

Bancroft was not a lone voice in his assessment of the situation. The 1894 report on the Columbia River salmon fisheries by the U.S. commissioner of fish and fisheries, Marshall McDonald, reviewed accounts of the decrease in salmon abundance in the upper Columbia River basin and concluded that in the twelve years between 1878 and 1890 salmon had become scarce in many areas of the basin. Both the causes and prognosis were obvious: "We must look to the great commercial fisheries prosecuted in the lower river for an explanation of this decrease, which portends inevitable disaster to these fisheries if the conditions which have brought it about are permitted to continue" (McDonald 1894, 5).

Concern was also voiced at the state level. Hollister D. McGuire, appointed in 1893 as Oregon's first fish and game protector, noted that the prized chinook were seriously depleted and that the high-volume fishery of the 1890s was maintained only by including smaller, less desirable fish, as well as coho and sockeye. The largest, most desirable chinooks now accounted for only half the salmon pack. McGuire was concerned that the Columbia River fishery would be irreparably harmed before conservation measures could bring the big chinooks back to their former abundance. Addressing himself to the state legislature in his reports, he cautioned that the industry itself was at risk.

As fish and game protector, McGuire recommended a series of measures essential for perpetuating Oregon's salmon. These measures read remarkably like those recommended previously for protection of Atlantic salmon in both Britain and New England. McGuire had little doubt about how to save Oregon's salmon (1896, 27–28):

- Replace the tangle of salmon fisheries laws with a single comprehensive law, coordinated with the State of Washington.
- Impose closed seasons based on run timing.
- Prohibit taking of salmon in spawning streams or near their mouths.
- Provide all dams with fishways.
- Build and fund the operation of hatcheries.
- Require screening of irrigation and drainage ditches to prevent fish from entering them.
- Appoint and support county fish and game wardens to aggressively enforce fisheries regulations.

To this day, the only measures that have been adopted and aggressively enforced are closed fishing seasons and support for hatcheries. Of salmon's four H's—harvest, hydropower (dams), habitat, hatcheries—the only H missing from McGuire's list was habitat degradation. Although no longer pristine, Oregon's rivers and streams were not yet under assault by the timber industry, agriculture, and urban development.

Concern over the decline in the Pacific salmon fishery was so great that in 1892 the U.S. Navy ordered Commander J. J. Brice to survey government installations for likely sites to set up salmon hatcheries, and to develop a plan to "restore salmon to their original numbers." Brice recommended establishing twenty-four hatcheries from California to Alaska. He also recommended setting aside the Klamath River basin as a fish preserve as a hedge against failure of the hatchery system. The hatcheries were built, but his recommendation for a salmon sanctuary languished.

Many of the fundamental causes contributing to the decline of Pacific salmon runs were quickly recognized by observers and government agencies, and regulation of salmon fishing on the West Coast had already begun before the rise of salmon canneries. Prior to 1877, when Washington instituted closed fishing seasons, regulations were intended to affect allocation of the catch rather than fishing intensity. Oregon imposed closed seasons the next year.

Unfortunately, however, enforcement was as lax as it was back East. Once again, convictions were rare and penalties light when violators were brought to trial in local courts. In 1903, H. G. van Dusen, Oregon's master fish warden, reported the widespread disregard for fishing regulations to the state legislature, asserting, "I found that fishing was being carried on in all directions and no pretense whatever being made to respect the law" (1903, 7).

Only two of the twelve men charged with violating the closed season in 1903 were convicted. They were fined $8 each, about the price of a case of canned sockeye.

Intensive fishing continued unabated and the Columbia River runs continued to decline. By the early 1920s, Washington's fishing fleet had grown to over fifty thousand fishermen, and eighty canneries were operating in the state. State Fish Commissioner L. H. Darwin warned, "No honest person, familiar with the facts, will attempt to deny that the fisheries of the State of Washington are being depleted" (quoted in Meeker 1921, 284).

By the start of the Second World War, the more commercially valuable species were particularly hard hit and biologists were advising that steps were urgently needed to ensure continued commercial salmon fishing. Writing before most of the dams were built on the Columbia River, Willis

Rich, a professor of biology at Stanford University and the first chief of research for the Oregon Fish Commission, warned of the dire consequences of further delay in addressing the causes of salmon declines.

> The future outlook for the salmon fishery of the Columbia River is not bright, but neither is it hopeless. . . . It will be necessary to restrict the commercial fishery to reduce catches, to improve conditions on the breeding grounds and to be eternally watchful that in the further development of water resources due consideration is given from the beginning to the needs of salmon conservation. . . . A few more years of inaction, and the runs may well have been reduced to a state of commercial extinction from which, if recovery is possible at all, it can be accomplished only after a long time and at great expense. (Rich 1940, 46)

Rich noted that the Columbia River chinook and sockeye runs were already dangerously depleted. He attributed the drop in sockeye abundance to habitat loss due to dams and flow diversions that had exterminated "many of the original races (or stocks) of this species." In Rich's view depletion of the Columbia River chinook was due primarily to overfishing.

> [T]he fishery during June and July is so intense that less than 20 percent of the Chinook salmon entering the Columbia River escape to the spawning ground and this involves no consideration of the fish taken in the ocean. . . . The continued effect of such over-fishing will undoubtedly be to still further reduce the abundance of these fish to a point where the total fishing effort will be curtailed. (45)

As Rich predicted, over time Columbia River salmon runs dropped to the point that many have no commercial value and some may actually disappear entirely.

The early history of the Puget Sound and British Columbia salmon fisheries parallels that of the Columbia River. The first salmon cannery on Puget Sound was built in 1877 at Muckilteo just north of Seattle. British Columbia's first cannery was built in 1871. By 1900 nineteen

Surveying the haul at Meyer's Packing Company, Seattle, 1895.

canneries were operating on Puget Sound. The fish traps that fed the canneries were so effective that concerned fisheries biologists began to argue that preventing wholesale extermination of salmon required allowing a guaranteed number of fish to escape the nets to reach their spawning grounds.

The efficient traps quickly outstripped the capacity of the canneries to pack the catch. Contemporary observers reported that in 1901 more Puget Sound sockeye were thrown away than made it into cans. Barge loads of excess salmon were towed into the Georgia Strait and dumped when canneries could not accommodate the haul. Cannery records therefore provide a bare minimum estimate of the total catch. Nonetheless, they illustrate trends over time. Production of canned salmon on Puget Sound increased slowly until 1890, when it began to grow rapidly, peaking just before the First World War. The cannery pack declined after the Great War, dropped even further during the

Salmon halfway to the rafters of a Pacific coast cannery.

Depression, rebounded substantially after the Second World War, and has progressively declined ever since.

Concerned over declining salmon runs, voters supported initiative after initiative to protect salmon. Roused by an aggressive advertising campaign in 1935, Washington voters passed Initiative 77 banning the use of salmon traps. But on the Columbia River things didn't really turn out as advertised. Oregon's fishing boats increased their catch once Washington's traps closed because of the new ban. A decade and a half after Initiative 77 the number of salmon reaching their spawning grounds in the Columbia River system had not increased. Instead, the attempt to regulate the type of fishing gear resulted in the transfer of much of the catch from one group of fishermen to another.

Just as technological advances revolutionized the transport of salmon to distant markets, new technologies also changed salmon fishing. By the 1930s, advances in marine technology allowed development of vessels designed for open-ocean salmon fishing. Huge floating canneries allowed harvesting and packaging salmon far from the rivers where runs originated. Moreover, by establishing a marine fishery, a country with no salmon, or one that had already depleted its own rivers, could harvest fish that originated in other countries. Rapid expansion of international fishing capacity began to hamper fledgling efforts to conserve Pacific salmon.

In particular, the Japanese salmon-fishing fleet expanded rapidly in the early twentieth century. Japanese fishermen began working Alaskan waters in 1904 and tensions steadily escalated over Japanese fishing in North American coastal waters. Just before the Second World War, the Japanese parliament authorized experimental open-ocean salmon fishing off the Alaskan coast. Japan's scientific investigation of the Bristol Bay salmon fishery in 1936 triggered a diplomatic crisis once it became clear that the "study" focused on catching and canning salmon in floating factories. Alarm bells went off in the U.S. fishing industry. The following year, a Japanese company approached cannery interests in Seattle about joint ventures to expand the use of open-sea canneries with Japanese equipment and labor. American labor and cannery interests were furious over this incursion into the North American fishery. Ultimately, the Department of State presented a statement to the Japanese government that concluded with a blunt warning to keep their hands off Alaskan salmon. American pressure convinced Japan to abandon its experiment, but friction over encroachment on Alaskan salmon frayed already strained relations between the two countries.

In reacting to the foreign threat to the salmon fishery, the U.S. government asserted priority over salmon originating in American rivers. As noted earlier, with ocean fishing there is no way of knowing where the catch originated. Out in the middle of the ocean it is impossible to know whether one is hauling in the last fish from an endangered run or the first fish from a healthy run. The American position held that overfishing offshore waters should not be allowed to undermine efforts to protect or restore salmon in rivers. The U.S. government further argued that the benefits of conservation efforts should accrue to those that bore the expense of such efforts in the first place. But as would be found later in the Atlantic salmon wars, there was no law governing fishing in international waters.

In the late thirties, Homer Gregory, a professor in the College of Economics and Business at the University of Washington, and staff researcher Kathleen Barnes undertook a study of the Pacific salmon fisheries for the American Council of the Institute of Pacific Relations. In *North Pacific Fisheries, with Special Reference to Alaska Salmon* they summarized the tension between technological advances in the ability to find and catch salmon and the need to restrain fishing intensity in order to sustain the fishery:

> From the native Indian salmon spear to a Japanese floating cannery is a long jump and represents a considerable advance in the technology of securing and preserving of salmon. In the course of the process, the existence of the salmon runs has been threatened as man has become more adept in exploiting the fish to his profit. To preserve the resources and to put it on the basis of perpetual returns has necessitated controlling this exploitation. It forces man to be moderate today that he may also profit tomorrow. (Gregory and Barnes 1939, 301)

The bombing of Pearl Harbor in December 1941 put negotiations over Alaskan salmon on hold.

As in the earlier global conflict, during the Second World War the government purchased most of the canned salmon produced from U.S. waters. But in contrast to the sound's increased salmon production during the First World War, Puget Sound salmon production decreased in the 1940s. The wartime labor shortage and competing demands on boats owing to the Japanese threat in the Pacific combined to reduce the capacity of the North American salmon fishing fleet. Salmon stocks started to rebound during the first lull in fishing intensity in decades.

The respite was brief. After the Second World War, the rapid pace of technological change continued to transform and intensify salmon fishing. In particular, the introduction of echo sounders and radiophones led to greater sophistication of open-water fishing. Advances in fishing gear also accelerated. Introduction of power reels to retrieve nets allowed bigger nets to be set. The size and power of boats, the number of boats, and their efficiency increased dramatically in the postwar years.

Fishing intensity doubled in the decade after the Second World War, renewing the pressure on dwindling salmon stocks.

Today, most of the American salmon fishery is located in Alaska, where both habitat and salmon runs remain relatively healthy. The first salmon cannery in the Territory of Alaska was established in 1878, and by 1888 the amount of salmon packed in thirty-seven Alaskan canneries grew to exceed the output of the Columbia River. The first federal regulations on salmon fishing in Alaska, passed in 1889, banned the obstruction of streams to migrating salmon and authorized appointment of federal agents to enforce the law and investigate the Alaskan salmon fishery. Subsequent acts in the late 1890s established a weekly closed period—a certain number of days per week when no fishing was allowed—and protected spawning grounds.

Early reports on the Alaskan salmon fishery describe fishing practices much like those that led to depletion of salmon elsewhere. There were plenty of fish to be had, plenty of fishermen having them, and no one trying to slow the take, as Gregory and Barnes explained: "There was no strong public opinion to back the enforcement of laws restricting overfishing. . . . Practices which experts and laymen alike are now unanimous in condemning were allowed to flourish without hindrance" (45).

Alaska's vast size confounded efforts to enforce early fisheries laws. The handful of fisheries agents had few resources at their disposal. Most critically, they had no boats of their own and relied on fishing boats or transportation supplied by canneries. Surprise inspections were impossible. Cannery interests could simply remove fish traps or barriers from streams before visits by federal agents.

Even so, early government agents stressed the continuing threat to the salmon fishery presented by industry practices. Fisheries Agent Joseph Murray, in a report to the U.S. House of Representatives, related that three or four corporations based in San Francisco were endangering Alaska's salmon runs. The companies routinely ignored closed times, completely blocked spawning streams with nets, and mined whole salmon runs until they were gone—practices that Murray noted had been outlawed for centuries in Scotland: "Paradoxical though it may appear, it is nevertheless true, that none are more anxious to save and perpetuate the salmon than the canners themselves, and yet their methods are such as, if continued, will very soon destroy them" (1898, 406). Other agents also reported that unfettered competition pushed the new Alaskan fishing industry to adopt unsustainable, short-term strategies. For example, Special Agent Howard M. Kutchin, in an 1898 report on the salmon fisheries of Alaska, wrote that "fierce competition, unrestrained by adequate laws, has been and is now operating to force packers, who can not help but realize the suicidal policy of such a course, into practices which they must know will surely be fatal to the permanence of their interests" (Kutchin 1898, 31). The federal government realized that without adequate protection Alaska's salmon runs would quickly follow those of New England and California down the road toward commercial extinction.

Alaska's being a territory rather than a state meant that the federal government retained broad authority to regulate fishing and land use. Federal regulation was seen as a necessary counterbalance to the natural incentives for individuals to collectively overexploit and deplete a limited, but renewable resource. On this point Homer Gregory and Kathleen Barnes observed in their study of Pacific fisheries, "A competitive industrial system. . . has made it virtually impossible for a brake to be applied on exploitation except by government control. Since the early days of the salmon industry the danger of unregulated exploitation has been apparent; yet any concern acting singlehandedly to protect the salmon would merely have sacrificed its own interests" (1939, 37).

Increased demand (and rising prices) for canned salmon during the First World War increased fishing intensity in Alaskan waters. By the end of the war 135 canneries were operating in Alaska. A year after the war's end, in 1919, special agent Dr. C. H. Gilbert and his assistant, Henry O'Malley, reported to the U.S. Commissioner of Fisheries that the Alaskan salmon fishery was at a crucial point. Unless a radically new administrative policy was established the supply of Alaskan salmon was threatened with eventual extinction. Gilbert and O'Malley reported that the high yield of the Alaskan salmon fishery was being maintained only by fishing operations' moving from areas that had been depleted into areas not yet fished. They proposed a 50 percent escapement rate (letting half the fish escape the nets) to ensure maintenance of the individual runs. The proposal to leave half the salmon in every river was a radical departure from prior salmon free-for-alls.

Concerned over the trajectory of the Alaskan salmon fishery and recognizing the need for action, the federal government developed assertive regulations to protect and foster Alaska's salmon. The most important of these new regulations was the 1924 White Act, which prohibited obstruction of streams and gave the secretary of commerce (and thereby the Bureau of Fisheries) the authority to establish closed seasons. In addition, the White Act endorsed the principle of 50 percent escapement, and provided for the development of regulations deemed advisable to achieve that goal. The act was a landmark shift toward incorporating biological considerations into regulation of salmon fisheries. It did not apply outside of Alaska.

Homer Gregory and Kathleen Barnes' study of the Pacific salmon fisheries contrasted the effectiveness of state versus federal salmon conservation efforts in the Pacific Northwest and Alaska:

> Judging by the record. . . it would seem that state conservation has not been as efficient as that of the Federal government. Particular difficulties hamper conservation in the states. Chief among these have been jurisdictional jealousies with special reference to Federal action, and the greater vulnerability of the state conservation policies to pressure from groups whose interests may be injured by regulatory action and whose influence counts more in state capitals than it does in the larger arena of national politics. (1939, 39)

Gregory and Barnes's conclusions still ring true today when state efforts tend to favor hatchery production in order to maximize short-term returns for commercial fishing interests.

Jurisdictional conflicts between state and federal agencies charged with regulating and protecting the salmon fisheries hampered conservation efforts. The problem of multiple jurisdictions plagued not only the Columbia River salmon, shared by Washington and Oregon, but also the sockeye runs of the Puget Sound and Fraser River, which supported fishing fleets from both Canada and the United States. Competition and cross-border differences in regulations began to undermine conservation efforts before the end of the nineteenth century. A joint Canadian-U.S. commission established in 1892 to investigate the condition and conservation of the region's salmon recommended application of the more conservative Canadian regulations to American harvest of Fraser River sockeye (which spawn in Canada). Uninterested in limiting their catch by applying more stringent regulations, American fishing interests increased the intensity of their fishing on Fraser River runs.

Canadian interests blamed American overfishing for the decline of Fraser River Sockeye. Americans retorted that the Canadians weren't really heeding their own regulations anyway. This stalemate led to the appointment of another commission to propose uniform regulations for the boundary fisheries of Washington and British Columbia. The commission proposed restrictions on the size and type of fishing gear, closed seasons, prohibitions against taking immature fish, and provisions addressed at preventing water pollution. But American approval of uniform fishing rules under an international treaty would put the federal government in the position of negotiating over issues that the state of Washington considered its prerogative. Although Canada passed legislation to adopt the new rules, Congress, under pressure from industry and state's rights advocates, failed to enact the rules. So the Canadians repealed their new rules and the process started all over again.

Decades later, John Cobb, in a 1930 report, "Pacific Salmon Fisheries," written for the U.S. commissioner of fisheries, described the dynamic of negotiations at this time between Canada and the United States:

Several abortive attempts have been made by the authorities of Canada and British Columbia on the one side and the State of Washington on the other to arrive at some equitable method of protecting this sockeye run. The former especially have professed an earnest desire to do something along this line, and there is no reason to doubt their sincerity. On the American side a few people, and among these a few of the more intelligent canners, pleaded for the enactment of laws that would adequately protect the salmon, but they were overborne by the great bulk of the packers and fishermen who, disregarding all the warnings and teachings of experience, insisted upon going ruthlessly forward with the slaughter, and when reproached with their short-sightedness clamored for the establishment of more salmon hatcheries, as though the latter could accomplish the miracle of increasing the supply of fry from a steadily decreasing supply of eggs. (Cobb (1930, 504)

The problem with short-term economic incentives is that they generally encourage individuals to overexploit resources without regard for long-term impacts, and this is not unique to salmon runs. But the fact that salmon pay no attention to international borders on their oceanic migrations further complicates management of the Pacific salmon. Alaskans catch salmon that spawn in Canadian rivers and Canadians catch salmon that spawn in Washington's rivers. Washington fish that swim north when they leave Puget Sound become part of the Canadian fishery once they enter Canadian waters. In the same way, Fraser River salmon that turn north become the target of the Alaskan fishery upon reaching Alaskan waters. So the Alaskans catch Canadian salmon, and the Canadians catch Washington salmon. Conflict over who catches whose fish kept derailing negotiations as runs kept declining and conservation measures became increasingly costly.

As salmon stocks in both the United States and Canada continued to decline, the two countries reached a draft treaty on limiting open-water, ocean interception (fishing) of salmon runs in 1982. Objecting to the potential need to reduce harvests, the state of Alaska and Puget Sound fishing interests blocked ratification of the treaty. Outraged Canadians threatened to block U.S. fishing vessels. Finally a compromise was reached and was ratified by Congress through the Pacific

Salmon Treaty Act of 1985; it committed both countries to reducing ocean interception of salmon without undue disruption of fisheries already in existence, and authorized creation of the Pacific Salmon Commission, with eight commissioners, four each from the United States and Canada to enforce the treaty. Though the treaty commits both countries to rebuilding salmon runs, it does not require the restraint of land-use practices that degrade habitat in spite of general consensus that such steps would be necessary to rebuild many depressed runs. Instead, the commission members argue over who can catch how many of the region's remaining salmon.

In addition to contributing directly to declining salmon numbers, overfishing also results in indirect effects that can impact the viability of salmon runs. Selective fishing of the largest fish reduces the size of salmon over time. Since smaller fish dig shallower nests, this creates a fundamental problem for fish adapted to burying their eggs just deep enough to avoid scour during winter storms. The reduced size of spawning salmon also means that some streams, or certain stretches of streams, that have coarse-gravel bottoms no longer provide suitable spawning habitat because the offspring of today's smaller salmon are vulnerable to being scoured out while incubating beneath streambeds.

Owing to favorable ocean conditions, larger-than-expected runs of salmon returned to the Columbia River in 2001. This could have been a year to allow stocks to begin rebuilding themselves. Instead, fishery managers decided to use the surprising returns to raise the allowable catch. Apparently, the system is set up such that when fewer fish come back than anticipated the harvest proceeds at the planned rate anyway, but when more fish than anticipated come back from the sea, the harvest increases. If bad years decrease fish numbers but good years do not rebuild runs to compensate for those bad years, the constant pressure on stocks ratchets the total population downward and runs can only decrease over time.

On the other hand, curtailing fishing pressure has proved to be an effective way to start rebuilding salmon runs in the past. A temporary ban on commercial salmon fishing can allow stocks to start to recover.

A famous example arose when the salmon catch on England's River Wye fell by two thirds between 1890 and 1901. The private proprietors of the fishery agreed to an arrangement to stop commercial fishing for five years, and then resumed with fewer nets and a shorter fishing season. By 1930 the catch grew to exceed the 1890 catch and was almost four times the 1901 catch. The short-term sacrifice reduced fishing intensity and provided the reprieve necessary to stabilize and rebuild the fishery.

But even temporary curtailment of fishing has been extremely difficult to achieve on publicly held rivers where overfishing was the obvious cause of declining salmon runs. New Brunswick's Miramichi River provides a case in point. In 1949, Percy Nobbs examined the state of the Miramichi salmon fishery for the Atlantic Salmon Association. Nobbs noted that the river lacked dams or natural obstructions that might impede fish passage, was unpolluted, and had extensive beds of spawning gravel (Nobbs 1949). Splash dams and log driving were virtually things of the past. Given the relatively good shape of the river, the reason that the salmon were still declining was obvious—too many fish were being caught and not enough were spawning. Nobbs's recognition that salmon restoration was as much a political problem as a biological one was implicit in his statement, "[S]hould it be decided to restore the Miramichi [salmon] it could be done in six years, but a lot of eggs would have to be broken to make that omelet" (23).

Nobbs noted how the government, instead of prioritizing the interests of future generations and making tough decisions, had between 1930 and 1939 solicited five reports on the state of the Miramichi salmon. Although all the reports came to similar conclusions, nothing was done. Politics tends to make commissioning reports easier than making omelets.

It has proved to be very difficult to restrict overfishing when the fish were viewed as public property available to all. Where fishing rights were privately held, as in Scotland, private interests could purchase rights and reduce impacts to sustain long-term productivity. Similarly, in federally controlled fisheries, such as Alaska's, the government could impose regulations tailored to protecting the resource. In most cases where local or state governments managed salmon fisheries, local interests applied substantial pressure to maintain the status quo, protect local people's immediate livelihoods, and avoid sacrifices necessary to sustain the long-term productivity of the resource. Left to their own devices, everyone wants all the fish that they can catch before someone else gets them.

In this situation, policy makers could be forgiven for looking for an easy way out that sidestepped the need to worry about stream conditions, tussle with federal lawmakers, or make difficult political decisions as to who gets to catch the fish. Hatcheries seemed to provide a painless solution to all of these dilemmas: Just make more salmon.

BETTER THAN NATURAL

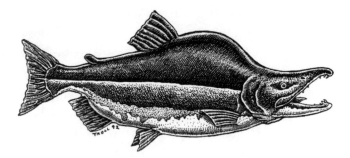

Nature cannot be conquered but by obeying her.

Sir Francis Bacon (1561–1626)

THE MEETING OF WASHINGTON'S INDEPENDENT SCIENCE Panel on February 28, 2001 was to be our first discussion of the role of hatcheries in the state's salmon recovery efforts. We met in Olympia on the fifth floor of the Natural Resources Building, and by 10:30 A.M. we had dispensed with housekeeping details and began discussing how fish hatcheries affect wild salmon. At 10:54 the building started to rock. Eyes darted about the room as we realized it was an earthquake. Xena sat up and stared wildly at me from under the conference table.

We were close to the epicenter of the biggest earthquake to hit Washington in thirty-five years. Stuff flew around the building. Filing cabinets flew open, discharged their contents, and fell over. Walls cracked. A completely unnecessary alarm warned that something was wrong and that everyone should leave the building. Less than a minute later when the building stopped shaking, everyone in the room had joined Xena under the conference table. The consensus among the ISP was that we would be wise to table the volatile issue of hatchery fish.

Sustaining salmon populations can be seen as either a narrow technical challenge or a broad ecological problem. In the first case, the "salmon factory" approach, hatcheries can be used to try and pump out more fish into the habitat so that one can keep fishing intensively. In the second case, the "wild salmon" approach, hatcheries can be used in conjunction with limited fishing and habitat restoration to let spawning runs recover—an approach based on fostering the viability of self-sustaining populations of wild salmon. Both the "salmon factory" and "wild salmon" approaches put salmon back in streams to be caught by commercial fishers. Hatcheries operated to rebuild wild runs release hatchery fish only as necessary, but also require reduced fishing intensity and protection of salmon habitat. Salmon factories, by contrast, rely on continual, ongoing releases, and the philosophy of managing hatcheries as salmon factories promised a painless way to treat the symptom of too few fish without curing the diseases of overfishing and environmental degradation.

Artificial cultivation of fish is an ancient art. Roman patricians maintained ponds in which sea-going (i.e., anadromous) fish could enter to spawn and then be captured on their return to the sea. After the fall of Rome, the art of anadromous fish culture declined in Europe. In medieval England fishponds were used to raise fish to supply monasteries and manors, and only freshwater and exotic fish such as carp were raised in these ponds. Practical status symbols for the elite, fishponds were not used for raising common fish like salmon.

Europeans rediscovered the cultivation of anadromous fish when the German Lieutenant S. L. Jacobi observed fish spawning in a river in the 1730s. Jacobi noted that the eggs were fertilized in the river, outside of the bodies of the fish. Intrigued, he produced live fish by collecting and mixing sperm and eggs, and incubating the developing embryos in boxes placed in streams. Jacobi experimented with fish culture for three

Squeezing milt from chinook male salmon to fertilize eggs at Big White Salmon Station in Washington State, ca. 1920.

decades before finally publishing his research. Acclaimed by the scientific community, his work remained an academic curiosity for almost a century until the general decline of European salmon prompted renewed interest in using hatcheries to rebuild and sustain fisheries in the face of widespread overfishing and river degradation.

Inspired by Jacobi's success, other scientists experimented with artificial propagation of European salmon. John Shaw, the Scottish gamekeeper for the duke of Buccleuch, began a series of experiments in 1833 to settle the longstanding argument over whether parr were indeed juvenile salmon. He collected fertilized salmon eggs from spawning beds and raised them until they grew into smolts. To ensure that the fish he raised came from salmon, Shaw stripped the eggs from a female salmon, artificially fertilized them, and again grew the resulting eggs into smolts, thereby settling the question beyond all doubt as to whether parr were juvenile salmon. In so doing, he introduced salmon hatcheries to Scotland.

Until the 1840s the practice of fish-culture in continental Europe was conducted primarily to investigate the natural history of salmon. This changed when two French fishermen, Joseph Remy and his friend Messieur Gehin, became frustrated by the declining salmon runs of the Moselle River. These two enterprising peasants studied the problem and began to collect salmon eggs from streams they fished, raising the eggs into fry, and releasing them back into the wild. Initial success led Remy and Gehin to set up a small-scale hatchery operation to capture adult fish and artificially fertilize their eggs. In 1850, the French minister of agriculture and commerce appointed Henri Milne-Edwards, a member of the French Academy of Sciences, to examine Remy and Gehin's progress in restocking portions of the Moselle River. Impressed with the report of a "veritable fish factory," the French government started to develop an extensive program to restock France's overfished rivers. The seductive idea of fish factories attracted the attention of scientists and governments across Europe. Remy and Gehin received the honor of an invitation to address the French Academy of Sciences and dine with President Louis Napoleon. Soon all of Europe was enthusiastic about the potential for hatcheries to restore the continent's decimated salmon runs. Few recognized that a critical reason for the restocking projects' success was that the habitat in the Moselle was still in pretty good shape.

By 1853, the proprietors of the Tay River fisheries had taken steps to start an experimental salmon hatchery near the River Tay to rebuild Scotland's salmon. After passage of the infamous 1828 act that added a month to the fishing season, the salmon fishery of the Tay had steadily lost value. By 1852 income from the Tay fisheries was reduced to half the revenue in 1828, despite the rising price of salmon. Lured by the promise of a restored fishery, the proprietors of the Tay salmon fisheries committed £500 to support artificial breeding. A committee set up to oversee the work built a hatchery at Stormontfield Ponds in the summer of 1853. The program received widespread attention because it succeeded in doubling the value of fish caught in the river within a decade.

William Brown, an angler who conducted and wrote about the Stormontfield experiments, argued that rearing salmon in a hatchery would greatly increase the return of adult fish. In particular, Brown held that by protecting parr from their natural predators, hatcheries would increase the survival of young hatchery salmon over that of wild fish. The great culling of salmon between deposition of fertilized eggs and their return as adults was seen as a needless waste of potential salmon. Brown lamented that less than one in several thousand fertilized salmon eggs returned to its river to spawn and become a marketable fish. Brown admitted that he did not know by how much the hatchery increased juvenile survival, but he was confident that increased salmon production from all those extra smolts could more than compensate for even very intensive fishing, stating: "There is little doubt but that the most of our salmon rivers (if well protected) would be sufficiently stocked with fish, if they were not, in every instance, over-fished by nets of all kinds; but when that is the case, artificial rearing, if generally adopted on a scale sufficiently large for the size of the river, would rescue enough of the ova from their enemies to meet this extra fishing" (1862, 36). It looked as though hatcheries could provide an easy answer to the problem of overfishing. The promise of increasing salmon runs without having to reduce fishing was irresistible.

Interest in salmon breeding grew rapidly. Scientists and fishery proprietors alike thought that by protecting salmon through their perilous early life stages the number of returning fish could be increased dramatically. Mankind had domesticated other wild game, like chickens and cows, so why not salmon? It seemed as though salmon could be shepherded through their youth and then set free to grow in the ocean and return home to be harvested.

Some of Brown's contemporaries, less easily seduced than he by the idea of creating salmon factories, believed that the real value of hatcheries lay in rebuilding wild runs. Among these was Andrew Young, another Scotsman, who discussed the artificial breeding of salmon in *The Natural History and Habits of the Salmon*. Young reported experiments confirming Shaw's earlier success in fertilizing salmon eggs and rearing salmon fry. Yet he expressed only guarded optimism about the future of salmon breeding: "Artificial breeding of fish has now become general in some rivers in Britain, as well as in France; but how far that system of producing fish will go to increase the numbers, or to keep up a national supply, has yet to be seen, for it is only yet on paper. However, with suitable laws and justice to salmon rivers, artificial breeding might, in course of time, become a supplementary fund to natural breeding" (Young 1854, 37). Thus began a split in philosophies of hatchery management, one that persists to this day.

During the wave of initial enthusiasm for salmon breeding the British government established its first hatchery in 1868 at Troutdale in Cumberland. Additional hatcheries soon spread across the British Isles. In some rivers the primary problem was simply overfishing, not compromised habitat, and in these healthy rivers restocking had some success. In one such place, Mr. Thomas Ashworth, the owner of the salmon fishery at Galway on the River Corrib in western Ireland, reported that for a small expenditure on hatching boxes he increased the value of his salmon runs twentyfold over the course of a few years. These facilities focused on rearing salmon fry and then releasing them into streams and rivers. Although hopes were high that hatchery juveniles would restock barren English rivers, the results did not live up to expectations for most rivers; as one writer noted dryly in the *Edinburgh Review*, "The sanguine predictions of teeming rivers and propagations, indefinite and infinite, of the Salmonidae, have not been verified" (1873, 153). British enthusiasm for reliance upon hatcheries soon faded as those hopes proved illusory for all but the few rivers where habitat remained productive and could support the smolts once they were released from the hatchery.

Hatcheries were also promoted in the New World. Richard Nettle, appointed superintendent of fisheries for Lower Canada Quebec in 1857, convinced the provincial governor to set up a salmon hatchery on the St. Charles River near Quebec. As catches of wild Canadian salmon continued to decline, officials embraced Nettle's enthusiasm for artificial propagation.

By the 1860s fisheries managers in Lower Canada also began suggesting that some streams in Ontario should be managed as preserves in which to rebuild salmon stocks. Pioneering fish culturist Samuel Wilmot convinced the government to give him control of a stream with a depleted run, where he banned fishing and built a hatchery to gradually rebuild the run. His success at restoring a wild salmon run fueled interest in hatcheries throughout North America, although jurisdictions unwilling to even temporarily suspend fishing or curtail further habitat loss handicapped subsequent efforts.

Canadian fishery authorities constructed a network of eight salmon hatcheries between 1868 and 1888. Combined with efforts to enforce fishing laws and construct fishways over dams, their program began to show results. The catch of New Brunswick and Nova Scotia salmon increased two- to threefold from 1867 to 1874. But over time the effects of pollution, logging, mining, and overfishing more than offset gains from hatcheries and conservation efforts. Canadian fisheries experts soon realized that even fish from a salmon factory needed rivers that could sustain salmon once they left the hatchery.

In contrast to early Canadian skepticism, south of the border hatcheries were seen as the salvation of the salmon fishery. The first commission of fisheries in the United States was established by the New Hampshire legislature in 1864 in response to concern over decimated river fisheries. Over the next two years other New England states joined in planning to restock the region's depleted rivers. Attempting to restock the Merrimack River, the commission procured a supply of eggs from a Canadian hatchery in New Brunswick. Though the first shipment produced just several hundred fry, it started the move toward reliance on hatchery-produced salmon in the United States. Then, in 1869, the Canadian hatchery stopped exporting eggs in response to local hostility to the idea of depleting Canadian salmon to rebuild American stocks. The commission of fisheries needed another source of eggs.

In 1870, the coalition of state commissions asked Charles Atkins, the commissioner from Maine, to locate a site for establishing the first American salmon hatchery. The next year, the commission signed a lease to operate a hatchery at Craig's Pond Brook near Orland, Maine.

By November 1871, the new facility produced an impressive 70,000 eggs, which were distributed mostly in Maine owing to the difficulty of transporting salmon eggs.

The United States Commission of Fish and Fisheries, formed in February 1871, was the first federal agency concerned with natural resource management. Its mission was to investigate the decline of foodfishes in U.S. waters and to evaluate methods for restoring and maintaining U.S. fisheries. The new agency moved quickly to support Atkins's work and relocated the hatchery closer to the mouth of the Penobscot River. At the peak of the stocking program, in 1874, more than 3 million eggs were shipped all over New England. Eggs were sent to Europe and even shipped west to Ohio, Michigan, and Wisconsin to try to extend the range of salmon. Local stocking programs appeared to be working as New England's salmon runs began to rebound.

Initially, hatchery operations were seen as a key element in a broad program to rebuild New England's decimated spawning runs. Writing in the popular periodical *Harper's New Monthly Magazine* in 1868, one observer noted, "[I]t will be necessary for each State to build and maintain the requisite fish-ways, stock the rivers, . . . and enact and enforce laws for the protection of the fish when ascending the rivers to spawn. Until this is done, and the wholesale slaughter of the parent fish . . . is prevented, we can not hope to have that plentiful supply, the want of which is now so much deplored" (Shanks 1868, 734). Meanwhile, with the emphasis on building hatcheries, provisions for maintaining fishways and preventing overfishing were implemented haphazardly, when they were implemented at all.

The U.S. Fish and Fisheries Commission also looked west to the Pacific salmon for help in reviving New England's rivers. The year after it was formed the commission hired Livingston Stone to set up a salmon hatchery in California. A retired, Harvard-educated clergyman with a passion for trout culture, his mission was to supply eggs for salmon restoration efforts in the Atlantic states. Stone traveled west with two young assistants and established his hatchery on the McCloud River, a tributary to California's Sacramento River. They designed and built the hatchery themselves and were operational within a year. They gathered and mixed eggs from adult chinook salmon, and the fertilized eggs were then packed in sphagnum moss, cooled by ice, and shipped east via stagecoach and railroad. Though most of the eggs produced in the first year were lost during shipping, the experiment demonstrated the ability of salmon eggs to survive shipping over long distances.

Demand was great for Stone's Pacific salmon eggs. Carefully packed crates of eggs were sent to restock rivers in England, continental Europe, and eastern Canada. Eggs were also sent across the equator to try and establish runs in Australia and New Zealand where salmon were desired by English colonists.

In many ways Stone's California sojourn was a great success. He introduced salmon to the Southern Hemisphere, and quickly became the reigning chess champion of California. But his efforts failed to achieve their primary goal. Attempts to use California salmon to rebuild New England's Atlantic salmon runs failed miserably, causing Stone to lament:

> I doubt if there was one person who had heard about it in America, whether interested in fish-culture or not, who did not believe that salmon were going to become abundant again in the Atlantic rivers on account of the introduction of the Pacific Coast fish. . . . That this did not prove to be the result was a stupendous surprise and disappointment. The eggs hatched out beautifully. The young fry, when deposited in the fresh-water streams seemed to thrive equally well. They grew rapidly and when the proper time came were observed to go down in vast numbers to the sea. What afterwards became of them will probably remain forever an unfathomable mystery. Except in very rare isolated instances, these millions of young salmon were never seen again. (Stone 1896, 218–19)

Stone had better luck using his eggs closer to home, and his hatchery received credit as runs on the Sacramento River began to recover. But the increasing runs generated a new round of cannery construction that absorbed the extra salmon produced by the hatchery. Eventually, overfishing and watershed disturbance depleted Stone's supply of eggs, and the hatchery closed down in 1884 (the historic hatchery now lies beneath the lake impounded behind Shasta Dam). Stone's efforts were nonetheless instrumental in selling hatcheries as the solution for the salmon crisis in the Pacific Northwest, and his influence led U.S. fisheries managers to accept William Brown's view that continual hatchery releases could substitute for maintenance of salmon habitat. Thus began an ongoing policy in the United States whereby hatcheries are used to try and support an aggressive commercial fishery rather than to rebuild self-sustaining spawning runs that could support more modest fishing without the expense of maintaining large hatcheries.

In the mid-1870s, less than a decade after Robert Hume set up his cannery on the Columbia River, concern over the long-term viability of the highly profitable salmon fishery led the Oregon legislature to solicit advice from Spencer Baird, a professor of natural history who was the first U.S. fish and fisheries commissioner. Baird identified the three primary threats to the salmon as overfishing, dams, and habitat degradation. He advised the legislature that, for a modest investment, hatcheries could produce enough salmon so that there would be no need to enact or enforce regulations to protect salmon rivers. In effect, Baird's view held that hatcheries could compensate for problems in the other three H's, habitat, hydropower, and harvest. Baird's idea that a productive salmon fishery did not require healthy rivers, wise stewardship of the land, or naturally reproducing spawning runs represented a further shift toward relying on salmon factories rather than addressing the effects of habitat degradation and overfishing.

Enthusiastic about prospects of producing more salmon without inconveniencing or curtailing their lucrative operations, a consortium of cannery owners enlisted Livingston Stone to set up a hatchery on the Columbia River under the umbrella of the Oregon and Washington Fish Propagating Company. The facility was built in 1877 on the Clackamas River and a decade later passed from its private builders into the hands of the newly organized Oregon Board of Fish Commissioners. The following year, unable (or unwilling) to fund its operations, the state deeded the facility over to the U.S. Fish and Fisheries Commission.

Failure of the state legislature to appropriate funds to operate the hatchery was just the beginning of bad luck. Though the facility initially produced a whopping 5 million salmon eggs a year, construction of mills and dams, together with impacts from logging, gradually destroyed the egg supply. A dam constructed downstream of the hatchery blocked salmon from returning, forcing the Fish and Game Protector to eventually sue (unsuccessfully) to enforce fish-passage laws and enable salmon to once again reach the hatchery. In a similar disaster, a hatchery built on the Siuslaw River never saw fish return from the sea because fishing nets captured the entire run. In the end, Stone's Mc-Cloud River hatchery, which was designed to provide eggs to recover Atlantic stocks, instead provided eggs to the Clackamas River hatchery to try to recover Columbia River salmon.

In 1880, Alvin. C. Anderson, British Columbia's inspector of fisheries, realized that the Pacific salmon are organized into separate local populations, with each river having its own distinct stock. He recognized that the supply of salmon in a river depended upon the number of spawners in that river. Anderson's views reflected earlier recognition that Atlantic salmon faithfully returned to their home stream. Acceptance of the stock concept led managers of British Columbia's salmon fishery to limit fishing effort by restricting both the timing of fishing and the type of gear permitted.

In the United States, the influential Livingston Stone maintained that salmon ran up rivers randomly, fostering the misconception that salmon were readily transplantable from river to river. Stone's rejection of the home-stream concept encouraged reliance on hatcheries, and transplanting of stocks became a cornerstone of salmon management in the United States. If salmon had no real dependence on their home stream, then why not move them around so as not to conflict with other desired uses of the land?

David Starr Jordan, the first president of Stanford University and the leading academic salmon biologist of his day, shared Stone's view, stating, "It is the prevailing impression that the Salmon have some special instinct which leads them to return to spawn on the same spawning grounds where they were originally hatched. We fail to find any evidence of this in the case of the Pacific Coast salmon, and we do not believe it to be true" (Jordan and Evermann 1902, 148). Jordan erroneously thought that young salmon that went to sea stayed within twenty to forty miles of their home river's mouth but then ran back up rivers randomly.

Canadians soon concluded that the Pacific salmon were no more able to be transplanted to a different stream than were the Atlantic salmon, and held to the home-stream theory for both species. They considered transplanting salmon from one stream into another more like a lottery than a sure thing and so shied away from introducing hatchery stocks from one river to rebuild depleted populations in other rivers. Because of their conviction that salmon were adapted to their home stream they considered hatcheries as a means to supplement depressed runs.

Stone and Jordan's ideas were influential, but other fisheries scientists argued that overreliance on hatcheries was unwise. In his official report for 1895, the U.S. commissioner of fish and fisheries, Marshall McDonald, noted that impassable dams reduce salmon production in proportion to the area blocked. He also recognized that hatcheries could not sustain salmon populations in the face of intensive unregulated fishing, and warned, "Until the States interested [in building more salmon hatcheries] adopt measures to restrain net fishing, so as to permit a portion at least of the salmon entering the river to pass up to their spawning-grounds, it is not deemed wise or expedient to attempt to increase or extend the work of artificial propagation of the salmon" (McDonald 1894, 18).

Commissioner McDonald believed that hatchery operations needed to be coordinated with state regulation of the fishery to ensure that operations were successful at not only producing fish for nets, but for rebuilding and sustaining wild runs. He also felt that "artificial propagation should be invoked as an aid and not as a substitute for reproduction under natural conditions" (1894, 4). The commissioner's conclusion that reliance on hatcheries to sustain salmon production was unwise did little to dampen American enthusiasm for salmon factories, which kept growing as salmon populations kept falling in the late nineteenth and early twentieth centuries.

In 1896, just two years after Commissioner Marshall McDonald's report, a U.S. Fish and Fisheries Commission agent, W. A. Wilcox, was asked during an interview with *The Morning Oregonian* about the future of the Columbia River salmon in light of the scarcity of salmon in New England. Mr. Wilcox avowed that the present generation knew little of the history of the decline of New England's salmon rivers, which once held salmon equal to the Columbia River's best. He attributed the decline of New England's salmon to pollution and considered it improbable that civilization could ever so contaminate the great Columbia as to endanger its salmon runs. He held that hatcheries could maintain a supply of salmon to match the intensity of fishing.

Voices calling for greater reliance on artificial propagation succeeded because hatcheries offered two major institutional advantages: They pumped cash and resources into facilities run by fish commissions; and they avoided the political problems associated with dealing with the known environmental causes of salmon declines. Proposing to simply produce more salmon was much easier than inconveniencing or interfering with powerful industries.

Hatcheries spread rapidly throughout the Pacific Northwest as a result of widespread recognition that salmon were overfished, and a lack of political will to restrict fishing. The first salmon hatcheries in the Puget Sound area were built around 1896. By 1900 five salmon hatcheries were operating in the northern part of Puget Sound alone. The hatchery built at the town of Bonneville in 1909 became the central facility of a network of hatcheries throughout the Columbia River system. Confident that hatcheries could sustain an intensive fishery even in the face of widespread habitat degradation, John M. Crawford, Washington State's superintendent of hatcheries, declared in 1911 that "there is absolutely no real reason for the eventual depletion of [salmon] by overfishing or the advance of civilization" (Smith 1979, 76).

Crawford's confidence appeared justified when Columbia River runs increased in 1913. Fishermen and agencies alike attributed the increase to the influence of hatcheries. Washington State Fisheries Director Leslie H. Darwin declared the Columbia River a restored stream.

Despite initial optimism, after a few more years it became apparent that the results of replenishing the runs on the Columbia from hatcheries were not so impressive. In hindsight, it appears that Mr. Darwin

Bonneville salmon hatchery, showing rearing ponds, ca. 1917.

gave hatcheries credit for the effects of a period of ocean conditions favorable to salmon returns. As runs began to decline again, better results were sought by shipping fertilized eggs to the central facility at Bonneville, feeding and raising fry, and then releasing them downstream below the influence of sawmills, dams, and irrigation ditches. Still, the millions of fry released into the Columbia every year had little apparent effect on the number of fish caught in the commercial fishery that the hatcheries were supposed to enhance.

Although enthusiasm for salmon factories remained high, evidence for a beneficial impact of hatcheries proved elusive. Scientists who looked into the issue remained skeptical, and Stanford University biology professor Willis Rich went so far as to assert, "In the early days . . . the hatcheries probably inflicted as much, or more, damage to the salmon runs as they did service of value" (1922, 68). By the late 1930s, when the potential effects on salmon fisheries of dams proposed for construction on the Columbia River became contentious, the opinions of early skeptics of hatchery programs were being repeated even in official reports on the potential effects of the dams:

[A]rtificial hatching has definite limitations. At best it is only a supplement for natural spawning. . . . [There are] requirements which have little, if anything, to do with artificial propagation, and cannot be managed by hatcheries. If we ignored these requirements, no matter how much we spent on building more hatcheries, the salmon fishery of the Columbia would be headed toward extinction. (U.S. Commissioner of Fisheries Report 1937, 60)

Still, hatchery success was measured by the number of fish released rather than by progress in rebuilding depleted wild runs. Salmon factories looked good when measured by the number of fry they produced—they didn't look as good if one looked at the number of spawners that returned. What hatchery boosters didn't understand was that after leaving home and hitting the real world, hatchery-raised fish survive at much lower rates than their wild cousins. Raising fish in a hatchery and releasing them to the wild may not increase the number of adult fish. Instead it simply rearranges when in their life cycle most of the fish will die.

In the wild, most die very young. Those few that survive are, on average, better suited for whatever life the local stream has to offer salmon. The dreadful culling of wild salmon in their early life stages equips the survivors for success on the rest of their epic journey out to sea and back home again. Charles Darwin called this natural selection.

Protected from day one, hatchery fish are not subject to this selective pressure. So when they are released into the wild, more of them are killed by predators or other natural hazards. Releasing hatchery fish into a stream is like dropping suburban teenagers into the middle of the Congo and asking them to walk out of the jungle to the coast. Few will make it. The hatchery fish that do make it back may be well suited for life in the marine environment, but the hidden price of reliance on hatchery fish is that resilience to disturbances, environmental change, and natural hazards in the equally crucial river environment may be bred out of a population.

Hatcheries treated fish production as a series of technical issues that, once solved, would allow production of as many fish as desired. First, the problem of hatchery design and basic procedures needed to be dealt with. Then fisheries scientists wrestled with the questions of whether to release eggs or fry; how large fry should be before release; what to feed the fish; what water temperature and pH were ideal; and how to time releases to maximize survival.

Today's technical challenges for hatcheries center on maintaining genetic diversity. Such concerns are leading hatchery programs to use native stocks. So we now have native stocks being used to prop up the hatchery stocks that were supposed to save the native stocks in the first place. Although the technical issues may have changed (nowadays focusing on how to maintain genetic diversity), the overall, production-oriented, technology-driven approach has not.

By the 1930s, the continued decline of salmon populations in spite of extensive hatchery programs led more and more biologists to recognize that hatcheries alone could not sustain salmon stocks. Indeed, John Cobb, dean of the College of Fisheries at the University of Washington, cited hatcheries as one of the principal threats to the salmon. Cobb's primary concern lay in the unquestioning optimism with which the hatchery programs were implemented. He cautioned that reliance on hatcheries to maintain salmon runs would eventually destroy the fishery.

In 1938, concern over declining salmon runs in Europe, the Atlantic states, and the Pacific Coast motivated the American Association for the Advancement of Science to convene a symposium on the migration and conservation of salmon at their annual meeting. One of the distinguished speakers at the symposium, Dr. Henry Ward, summarized efforts to transplant and introduce new runs of Pacific salmon, stating: "A few of these experiments have been successful in a degree but none of them in a large way. On the other hand, most of them have been total failures and these include experiments that were large and were carried out by able, energetic and well trained personnel" (1939, 64).

Part of the problem in devising workable plans and policies to protect salmon fisheries was that as yet there still was no consensus on the home-stream theory, on the extent of adaptation of salmon runs to conditions in their home stream. Professor Willis Rich summarized the available evidence and concluded that salmon conservation measures need to be based on the home-stream theory:

> [I]t is obvious that the conservation of the species as a whole resolves into the conservation of every one of the component groups. . . . Diverse evidence points so clearly to the existence of local, self-perpetuating populations in the Pacific salmon that . . . practical conservation measures must be based upon the acceptance of the "home stream theory" as an essentially correct statement. (1939, 45, 47)

Though, as Rich noted, the need to base hatchery programs on local stocks was clear, he also saw that state fisheries managers did little to protect wild runs that provided the genetic bank from which to draw robust local broodstock, commenting that

> about the only protection given to the Columbia River salmon has been that afforded by artificial propagation. Biologists in general are skeptical of the claims made for artificial propagation, and rightly so because these claims have often been extravagant and the proof is entirely inadequate. Indeed, many conservationists feel that the complacent confidence felt by fishermen, laymen, and administrators in the ability of artificial propagation to counterbalance any inroads that man may make upon the supply of a propagated species is a serious stumbling block in the way of the development of proper conservation programs. (1941, 429)

As the limitations of reliance on salmon factories became apparent, Canadian fisheries managers started to deemphasize hatchery programs in regions where habitat remained in good condition. In the 1930s Canada abandoned its policy of relying on hatcheries for salmon production after government scientists determined that they were not a viable substitute for natural propagation and were expensive and unable to sustain intensive fishing over the long run. Hatcheries were to be used only in cases where adequate reproduction could not be sustained naturally. The last of British Columbia's early hatcheries closed in 1937.

By 1939 Alaska also abandoned its initial hatchery programs. Advocated early on in Alaska, hatcheries lost their appeal after the White Act established the principle of 50 percent escapement, letting half the fish escape the fishing nets. Where there was more or less natural habitat and half the runs made it back to the spawning grounds, hatcheries were seen to be impractical affairs whose costs outweighed the benefits they produced. Over time, however, hatchery programs were renewed in many areas as a result of pressure to embrace the seductive promise of generating more fish without restoring rivers. But because Alaska possesses large amounts of unpolluted, accessible habitat and fishing is well regulated by federal authorities, Alaskan hatcheries have not been as damaging to wild runs as were state-run salmon factories in the Pacific Northwest.

Through the 1970s and 1980s agencies charged with fishery management in the continental United States continued to advocate reliance on running hatcheries as salmon factories to maintain salmon populations. At a National Marine Fisheries Service workshop held to discuss how to protect salmon under the Endangered Species Act in 1978, a service biologist maintained that the "release of smolts throughout the watershed . . . should ultimately provide a reasonable substitute for the original wild stocks" (Cone and Ridlington 1996, 211). John Cobb's fears about hubris turned out to be well founded. Even with aggressive hatchery programs, salmon runs continued to decline in Oregon, Washington, and Idaho. Hatcheries intended to bolster and supplement natural fish production ended up contributing to the decline of wild salmon.

In 1896, Washington state hatcheries produced 4.5 million chinook fry; in 1950, they released 28.9 million; by 1968 they were pumping out 92.7 million. But the harvest in the commercial fishery wasn't increasing. State fisheries managers knew they were replacing wild fish with hatchery fish but saw this as a way to sustain a fishery in the face of regional development and consequent habitat degradation. In its 1968 annual report, the Washington State Fisheries Department acknowledged this state of affairs, stating, "It is obvious that since total catch has not changed a great deal these past years that the contribution from naturally produced stocks of chinook and coho have been very adversely affected by the inroads of 'civilization,' and that the up-take has been borne by an increased catch of hatchery propagated stocks" (102).

Hatchery operations on the Columbia River released 100 million to 120 million juvenile salmon per year in the late 1980s and early 1990s. During this period less than half a million adult salmon returned to the hatcheries each year. This net return of less than 1 percent of the fish that went to sea was not all that much better than the survival rate for wild Pacific salmon.

The emphasis on hatcheries for sustaining Columbia River salmon led to a major reorganization of the geography of salmon production. Over 95 percent of the hatchery fish were released below Bonneville Dam in the lower Columbia River, whereas most of the losses from impacts of the dam system was on runs in the upper Columbia basin. Hatchery managers on the lower Columbia favored releases of fall-run chinook and coho, the species that supported the commercial ocean fishery. In the upper Snake River, hatcheries favored coho and steelhead (an anadromous trout), which sustained the sport fishery. The depleted sockeye and summer chinook runs above the dams were not supplemented by hatcheries, continued to dwindle, and are now listed as endangered on the Snake River. Over time, the policy choices concerning which species of fish to supplement and where contributed to the decline of the most vulnerable runs of the Columbia River basin.

Wild stocks fell fastest in streams that received the most attention from hatchery managers; for example, large hatchery releases of coho on the lower Columbia River accelerated the decline of wild stocks. Although less fit for survival overall, hatchery fish are quite aggressive when it comes to feeding. Once released into a stream, hatchery fish compete with wild fish for the same food. It is not really a fair fight. Hatchery fish are used to fighting other fish for access to food, and are more aggressive than their wild cousins. Hatchery fish also grow bigger faster on the hatchery diet and tend to be larger than wild fish. Hatchery salmon even eat smaller wild salmon. So they are not just bullies, they are big cannibalistic bullies. Efforts in Oregon to use hatchery fish to rebuild coho populations had the unintended result of simply replacing remnant wild populations with hatchery stocks.

Management of fisheries with a large component of hatchery fish also contributes to overexploitation of wild stocks, because the fishing intensity level that is calibrated to harvest a large proportion of a hatchery population disproportionately impacts small wild populations. Because wild salmon are caught in the same nets that seek the hatchery fish, highly productive hatcheries can ensure overexploitation of remnant natural stocks.

Hatchery-produced fish also damage the gene pools of the wild stocks among which they are released, or into which they accidentally find their way. One study in Oregon found that three quarters of the coho salmon spawning in the Yaquina River were hatchery fish, probably escapees from a fish farm in the estuary. Wild salmon have a great deal of genetic diversity, with each run adapted to its home stream. But hatchery stocks typically come from a small sample of an original population and thus exhibit a comparatively impoverished genetic variability. Even a modest number of wandering or escaped hatchery fish can genetically swamp a small wild population, thereby reducing its genetic diversity and increasing its vulnerability to environmental change.

Thus, adoption of an aggressive hatchery program frequently does not increase the total number of salmon over the long run. Instead, hatchery populations dependent upon human intervention gradually replace naturally spawning wild populations.

Hatchery stocks also introduce diseases into wild populations. Transplants from a Swedish hatchery stock in 1975 wiped out rather than restored Atlantic salmon runs in Norwegian rivers by introducing a parasite (*Gyrodactylus salaris*) that killed most smolts. Through hatchery transplants the parasite spread to rivers around the Baltic Sea, decimating or destroying runs in dozens of rivers in Finland, Russia, and Sweden. Another unfortunate example of a transplant gone wrong, also from the Baltic, is the case of M–74, the name of a disease that destroys 80 to 90 percent of wild stocks. Although hatchery stocks are also susceptible, M–74 syndrome can be treated by immersing eggs and fry in a thiamine solution. There is no practical way to treat wild fish. After rampant overfishing and hatchery programs reduced the genetic diversity and therefore resilience of wild fish in the Baltic, M–74 may well finish off the remaining wild Baltic Sea salmon. Such horrific unintended consequences of introducing hatchery fish from one area into another illustrate the risks inherent to viewing salmon recovery as a technical challenge rather than an ecological problem.

A comprehensive 1990 analysis of hatchery programs conducted by William H. Miller for the Bonneville Power Administration (BPA) concluded that hatcheries had successfully produced fish for sport, tribal, and commercial fisheries. But Miller went on to conclude that merely 8 percent of the more than three hundred projects reviewed were effective at supplementing wild runs. He also was troubled by the effects of hatchery fish on wild populations: "Examples of success at rebuilding self-sustaining anadromous fish runs with hatchery fish are scarce. . . . Adverse impacts to wild stocks have been shown or postulated for about every type of hatchery fish introduction where the intent was to rebuild runs" (1990, iii).

After relying on hatcheries for decades, BPA began to realize that the approach did not substitute for accessible, hospitable habitat. Miller, the author of the BPA-solicited report, stated, "Overall, we concluded that protection and nurturing of wild/natural runs needs to be a top management priority. There are no guarantees that hatchery supplementation can replace or consistently augment natural production" (iv).

The bottom line on hatcheries is that throughout the Pacific Northwest, salmon (both hatchery and natural) have continued to decline even though hatcheries have spent millions of dollars to produce hundreds of millions of fry. Oregon's hatcheries released tens of millions of salmon in 1985. And yet Oregon's salmon are in trouble—its wild ones in serious trouble. After more than a century of hatchery supplementation, Columbia River salmon runs remain at less than 10 percent of their historic size. Wild spawning populations account for only a few percent of their former abundance. Not only did hatcheries fail to revive the runs or sustain the fishery, but now most of the fish come from hatcheries.

Hatcheries are also expensive. In the early 1990s Oregon's hatchery system cost about $15 million annually. The cost to produce a single adult salmon ranged from ten to twenty-five dollars for most runs, to over a hundred dollars per fish for some stocks. Over time, hatcheries came to produce fewer and fewer fish at a cost that threatened to exceed their commercial value. Hatchery fish are neither an ecological nor an economic bargain.

In December 1998, the Scientific Review Team of the Northwest Power Planning Council reported on its review of science and impacts related to artificial production of salmon, and compared their review to three previous scientific reviews. There was uniform agreement on most points. Hatcheries failed to mitigate the effects of habitat loss and damage in the Columbia River basin, in no small measure because hatchery practices did not take into account the biological diversity of salmon and the role of environmental factors in their life history.

The Scientific Review Team found that a radically different approach was needed if hatcheries were to have a positive role in salmon conservation. Because of the need to maintain adaptability to future environmental changes, the new approach would need to minimize impacts on natural populations and preserve the genetic structure and diversity in salmon. And the new approach had to work on the premise that hatchery production is not independent of natural systems. The success of a hatchery program depends on the fitness of the stock, the quality and constraints of the natural habitat, and how well hatchery production is integrated with the natural ecosystem. In essence, the team endorsed the home-stream theory— the need to manage populations on a stream-by-stream basis, with the understanding that hatcheries could be used to supplement but not to replace reproduction of wild salmon. Andrew Young's vision for the role of hatcheries in salmon management weathered the test of time better than William Brown's idea of using hatcheries as salmon factories.

A major problem for implementing Young's philosophy is that many rivers can no longer foster the survival and growth of salmon, let alone their reproduction. Hatcheries also require a continual flow of money to sustain salmon production. Wild runs are a good deal for those who harvest them—like cattle grazing for free on public lands. However, reliance on hatcheries for salmon production is vulnerable to economic or political pressures that can lead to closure or defunding of facilities or hatchery operations. There is no evidence that, economically or ecologically, hatchery-based fisheries can be sustained over the long run.

Perhaps the most dangerous aspect of the historic reliance on hatchery production to sustain salmon populations is that the system has created the illusion that hatcheries can make up for the environmental changes and overfishing that led to declining salmon runs in the first place—even though the accumulation of evidence should destroy this illusion. The public and policy makers have been deceived into believing that we can sustain production of a valuable, renewable, and culturally important resource while simultaneously degrading the environment and the conditions upon which that production depends. John Cobb's and Willis Rich's insights into the dangers of over-optimism about hatcheries remain as relevant today as when first advanced more than half a century ago. Still, hatcheries remain wildly popular in most areas. Who, after all, could oppose making more salmon?

Fish farms are the ultimate "salmon factory," installations that grow salmon in submerged cages. With no need for habitat, salmon are raised in pens, starting in freshwater then moving to the sea or estuaries to mature. Forget the dangerous trip down the river to the ocean and back. Salmon farms keep the slippery devils under lock and key their entire lives, raising fry into adults as captives in open-water pens. Fish farming is seen by some as a major threat to wild salmon. Others see salmon farms as the salvation of wild stocks through reduced commercial fishing pressure. Whether blessing or curse, salmon farming is growing fast.

Norwegians pioneered Atlantic salmon farming in the 1960s. The idea spread to Scotland in the 1970s, and reached Maine in the 1980s. Production in Maine grew from nothing in 1986 to 16,400 metric tons in 2000. Between the late 1980s and early 1990s salmon farming in tidal rivers and along the North Atlantic coasts of Europe and North America grew tenfold from 30,000 to 320,000 tons per year. Globalization of salmon production and consumption spawned major salmon-farming operations not just in the North Atlantic, but also in Chile, South Africa, Australia, and New Zealand. By 2000, global production of farm-raised salmon approached a million metric tons.

Salmon farming proved so successful that oversupply drove prices down to a point where commercial fishing for wild salmon is no longer viable in some areas. By the 1990s farmed salmon dominated the Atlantic salmon market in the United States and Canada. European salmon farms now produce almost 100 times the catch of wild European salmon. In some river systems more than nine out of ten salmon are farm-raised.

Little is known about the long-term environmental effects of massive salmon farming operations other than that some farm fish escape, and the farms introduce concentrated wastes and diseases into rivers and estuaries. But it is known that wild stocks in Scottish rivers declined as fish farming grew. Conservation interests in Scotland and Ireland now fear that high concentrations of sea lice introduced by farm fish could wipe out wild smolts as they swim downstream past salmon farms. Wild stocks in Norway also were hard hit by disease and crossbreeding with escaped farm-raised fish.

The ocean is a hard place to maintain a fence. Each year as many as 10 percent of the Atlantic salmon in fish farms escape due to storms, damage to their pens, and from fish spilled during transfer to or from pens. These fish are not imprinted with any particular home stream, having been raised in estuaries or coastal waters, and head out to sea and then return to explore nearby rivers where they intermingle and interbreed with local salmon. Hundreds of thousands of farm salmon have escaped into Scottish and Norwegian rivers. In some years escapees account for a majority of spawners in some rivers.

Salmon farms in eastern North America have also had problems keeping their stock confined. Salmon escape from the six hundred floating net-pens in Maine's coastal waters at all life stages. A National Research Council committee estimated that 180,000 fish escape from salmon farms in Maine each year—about one hundred times the number of wild Atlantic salmon left in New England. In some years, salmon that escape from farms represent the entire spawning run in some rivers. Although the NRC report found that wild fish outcompete hatchery and farm fish, the committee was concerned that wild salmon's adaptive advantage might not prove sufficient to overcome the huge number of fish farm escapees.

By the 1980s, when New England's fish farms were just starting, growing concern in Europe focused on the eclipse and potential extinction of indigenous salmon stocks by fish-farm escapees. Norway established a salmon sperm bank and began to monitor the extent of hatchery and farm-bred fish in sixty-two rivers. The finding that in some rivers only one in five of the returning adult fish were from wild stocks elevated fears over the loss of genetic integrity of wild salmon. Some started to call for the use of sterile salmon in fish farms, but the salmon- farming industry steadfastly denied that hatchery fish would interbreed with wild fish. Despite such assurances, by the late 1980s biologists had documented farmed fish of both sexes spawning with wild salmon. The threat was real.

Atlantic salmon farming has even spread to the Pacific Northwest. Regulators swallowed industry assurances that Atlantic salmon would not escape into Pacific waters. But escape they have. It apparently came as a further surprise when escaped Atlantic salmon were found spawning. Regulators continue to trust industry assurances that Atlantic salmon cannot survive and reproduce in the wilds of western North America.

A recent experience in Asia brought both the decline in wild salmon runs and the globalization of salmon farming into sharp focus for me. Every four years the International Geomorphology Conference is held in a different country. Geomorphologists from all over the world gather to discuss rivers, landslides, volcanic eruptions, and landscape evolution. A special focus of the 2001 meeting in Tokyo was river restoration and the interaction of geomorphological and biological processes. Brian Collins and I traveled across the Pacific to present our research on historical changes to Puget Sound rivers. I also thought we might learn something about Japanese salmon.

Sitting on the train into Tokyo from Narita Airport and gazing out the window, we watched the lush Japanese countryside flash by. But I felt as though we'd rolled onto the set of a science fiction movie upon entering the city, a sea of concrete draped by flashing neon. Tokyo is a landscape of tight angular spaces, of wild colors splashed across gray backdrops.

Our first day in Tokyo Brian and I got up early, eager to explore the city. At the top of our list of places to visit was the central fish market, where I expected to see salmon ready to be sushi-ized. Hopping two subway trains, and navigating a maze of small streets, we arrived at our goal.

The Tokyo fish market is an immense tomb—a city of dead fish. Seeing this necropolis, it was easy to understand why the seas are emptying. Five thousand tons of fish pass through the market each day. We walked down aisle after aisle of squid, urchins, tuna, and unidentifiable marine life. Gawking at modern samurai wielding swords to turn huge tuna into sashimi, we walked on and on through a maze of fresh and frozen fish. Past the mountain of Styrofoam boxes being pushed into a 20-foot-tall pile by a bulldozer. But still no salmon.

Finally, atop a box flying a Norwegian flag, we spied an Atlantic salmon, a cross-hemisphere traveler limp in a Styrofoam crypt. In an hour wandering through the Tomb of the Seas we saw only a few boxes of Pacific salmon. Half the salmon were from Norway; most of the rest were Chilean. Atlantic salmon raised in pens and shipped halfway around the world supplies Tokyo's insatiable daily appetite for fish. Leaving the market on the way back to the subway we stopped next to a gentleman on a bicycle stopped at a traffic light. On the back of his bike was a box of salmon bearing the label of the Olympic Fish Company, LaConnor Washington. Japanese salmon were nowhere to be seen.

Japanese salmon are virtually gone except for remnant runs in northern Hokkaido. The ability to ship and market salmon all over the world means that Norwegian, American, and Chilean salmon now satiate the Japanese appetite for salmon.

Without effective regional checks against overexploitation of wild stocks, the day may come when there will be no more salmon outside of farms. Are salmon the next chicken or cow, to be domesticated and vanish from the wild?

Salmon farmers can choose the color of the fish they raise through food additives that produce a healthy-looking red hue. Raised in a pen and fed salmon chow, pellets made of compressed sardines, anchovies, and other fish, and which lack the krill that colors wild salmon, the flesh of farmed salmon would be an unappetizing gray. So the industry adds pigments (canthaxanthin and astaxanthin) found in flamingos, lobsters, and shrimp to produce vibrant colors. The Swiss pharmaceutical company Hoffmann-La Roche produced a clever device to help fish farmers choose just the right shade for their salmon. Called a Salmofan, it consists of a folding sample chart that allows a fish farmer to choose the color of salmon just as one chooses paint.

Soon salmon farmers may be able to stock their pens with fish genetically modified to grow four times as fast as wild salmon. Fish enhanced with two extra genes to produce growth hormones year round would cut the cost of producing salmon and could help reduce both fishing pressure on remaining wild stocks and the potential for fish farms to introduce disease and parasites to wild salmon by cutting the time salmon spend in crowded pens.

But critics fear that transgenic salmon, which they call "Frankenfish" because the growth-inducing genes come from other species of fish, could wreak havoc on wild salmon by either competing with wild fish or through the Trojan gene effect, in which larger transgenic males attract a large share of mates but produce less fit offspring. Concerned over the potential risks to wild salmon, the Washington Fish and Wildlife Commission banned the cultivation of genetically engineered salmon in December 2002.

Advances in biotechnology may mean that we can continue to eat salmon without needing the fish itself. Scientists have devised ways to grow fish meat in the laboratory. A lump of fish meat used as a starter (sort of like for sourdough bread) is bathed in fats that become assembled onto the original lump. The original impetus for this research was to supply a source of high-protein food for long space flights. The ability to grow fish meat without the bother of caring for (and disposing of) the fish raises new issues and concerns for salmon. Some may ask why we couldn't just bank samples of the best fish, stick them in the deep freeze, and grow pieces as we want them? Why not just preserve a sample of their DNA, or a map of their genetic blueprint so we could recreate them from scratch in the kitchen of the future? While this day remains in the realm of science fiction, it is conceivable that biotechnology will remove the need to have salmon in order to eat salmon. That can't be good news for wild salmon.

The increase in salmon farming has driven the price of salmon down, but wild salmon are still worth more than fish raised in a cage. They simply taste better. Will the rise of industrial salmon farms reduce the pressure on wild salmon? Or will they contribute to the environmental stress on dwindling wild stocks? The answer depends in part on how we manage our rivers. Salmon not only need rivers, they need rivers with certain traits. As recognized for a thousand years, foremost among these traits is unimpeded access to the sea.

POWER FOR THE PEOPLE

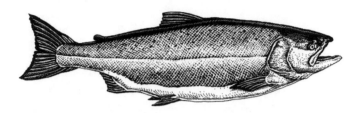

Roll along, Columbia,
You can ramble to the sea.
But river, while you're rambling
You can do some work for me.

Woody Guthrie, "Roll, Columbia," 1941

IN 1992, A SOCKEYE SALMON NAMED LARRY SWAM FROM the sea heading home to Idaho's Redfish Lake, up and over fish ladders at thirteen dams without eating. Like his ancestors, Larry traveled upstream at an average rate of 1 to 2 miles an hour, hugging the banks in water generally less than 30 feet deep. In the 1890s thousands of sockeye made this journey up the Columbia and then the Snake river to return to Redfish Lake.

Lonesome Larry was the only sockeye that made it back to Redfish Lake in 1992. Since he came alone, the National Marine Fisheries Service had a surprise waiting for him. Instead of a nice female to spawn with, he was clubbed on the head and had his sperm squeezed out of him by technicians to fertilize hatchery fish from which the agency hoped to revive his kind. Larry himself, the last wild Redfish Lake sockeye salmon, ended up as a display on the wall of a museum managed by a dam construction company that built many of the dams Larry swam over.

In 2002, on our way from Seattle to a writing retreat in Boise, Xena and I paralleled Larry's thousand-mile trip up the Columbia and Snake rivers. We drove it in a day, stopping for lunch along the way.

Trapped in the rain shadow of the Cascade Range, the Columbia Plateau of eastern Washington is dry, really dry. But unlike many deserts, it is fertile because of the wind-delivered silt that covers much of the plateau. You just need to add water to grow a lot of crops, and now, dams and irrigation ditches tap eastern Washington's big rivers (the Columbia and its major tributary the Snake) to water an ocean of grain in the middle of an arid plateau. More than half a century since the first dam was built, the irrigation system that transformed the high desert of eastern Washington into a productive breadbasket has become an untouchable regional entitlement.

The Columbia River was the logical physiographic boundary between the United States and British territories in the Pacific Northwest. In fact, the Hudson's Bay Company even established a British presence north of the Columbia River at Fort Nisqually prior to American settlement. The location of the border remained contentious until 1846, when the British relinquished their claim to what is now Washington State. Why did the British yield this area so easily?

One story holds that Washington is U.S. territory because Columbia River salmon would not take a fly. The brother of Sir Robert Peel, the British prime minister, was serving in the British Navy at Victoria, British Columbia, at the time of the settlement. Despite repeated attempts, this gentleman at the edge of the empire apparently had no luck in fly-fishing on excursions up the Columbia. Writing to his brother the prime minister, he is said to have advised him that that Columbia River salmon were too stupid to take a fly and that the river "wasn't worth a damn."

Named after the first ship to sail up the river in 1792, the Columbia is the fourth largest river in the continental United States (behind the Mississippi, Colorado, and Rio Grande). Draining 260,000 square miles of temperate rainforest, mountains, and high desert, with headwaters in Canada and Idaho, the Columbia River crosses through arid eastern Washington and skirts around the Cascade Range to head for the coast through a narrow gorge at The Dalles, just east of Portland, Oregon. The Columbia is an ancient river, older than the Cascade Range, having cut its way across the rising mountains millions of years ago before the range grew to its present height. The river is older than the Pacific salmon.

The west and east slopes of the Cascades are different worlds. Winds coming off the Pacific Ocean deliver the rains responsible for the Pacific Northwest's soggy reputation. But the rain stops at the Cascades, wrung from the winds by the high peaks along the crest of the range.

Twenty million years ago, before the Cascades rose, eastern Washington was forested, much like western Washington. The rising mountains formed a topographic barrier that cast a growing rain shadow across eastern Washington. The forests east of the rising range wilted away as the Columbia Plateau dried out, so that today Ponderosa pine forest on the east slope of the range yields to sparse scrub on the Columbia Plateau. Incised in a canyon cut slowly over 15 million years, the Columbia River hoards its water as it flows past the uplands without sharing a drop.

The free-flowing Columbia had its own seasonal rhythm. Spring snowmelt turned the river into a raging torrent. David Douglas, a Scottish botanist who canoed up the river in June 1825 (and for whom Douglas fir is named), described the Columbia in the middle of the snowmelt season as exceedingly powerful, making upstream travel slow and arduous. The river level varied so much through the year that dramatic features such as Celilo Falls were exposed only in late summer and early fall—the rest of the year they were submerged beneath rapids. Today you need scuba gear to find the location of the old falls beneath water impounded behind Bonneville dam. Other than between the sea and the lowest dam, the last free-flowing stretch of the

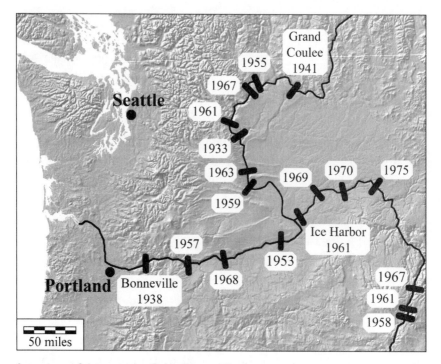

Locations of dams on the Columbia and Snake rivers.

Columbia runs through the Hanford nuclear reservation. The modern Columbia is more like a long string of lakes than a river.

Was damming the Columbia a good deal? Naturally, the answer depends on whom you ask. To the farmers on the irrigated plateau it was a great deal. Cheap power produced by the turbines at the dams and water diverted from the river still fuel the economy of eastern Washington and much of Oregon and Idaho. It was a disaster for the Indians and the salmon. Celilo Falls, where people had fished for over eight thousand years, slipped beneath the water. As more and more dams were built, the salmon that provided the livelihood for the original inhabitants of the Columbia River basin began to disappear.

Settlers who came to the West Coast knew of the role of dams in decimating salmon runs in New England and Europe. There was no desire to repeat these stories. In 1848, the Oregon territorial constitution proclaimed that the rivers and streams "in which salmon are found or to which they resort shall not be obstructed by dams or otherwise, unless such dams or obstructions are so constructed as to allow salmon to pass freely."

Meeting in its first session a few years later, the newly formed Washington State Legislature also passed laws making it illegal for anyone to block salmon from running up a river or stream. Though the intent of such laws seems clear, they were generally ignored or circumvented in short order.

In the early 1900s a large number of dams were built to supply water and power for western Washington's growing population. The city of Seattle, for one, dammed the Cedar River in 1901 to provide water for the growing metropolis. Built without provision for fish passage, the dam stood in unchallenged violation of state law for over a century. Other dams soon followed to provide water and power. Enforcement of fish-passage laws was lax as the Washington Department of Fisheries kept busy regulating the expanding fishing fleet and building an extensive network of hatcheries.

As dams started to proliferate across the region, fishing interests voiced concern about the potential impact on the salmon fishery and called for enforcement of fish passage laws. In 1903, a critic sounded the alarm in the *Pacific Fisherman*:

> The matter of harnessing the waters of our rivers by the immense dams that are being built across them for power, irrigation and milling purposes is sure to jeopardize the fishing industry of this state; for, as a general thing, they are building dams across the most desirable streams. . . . These are almost complete barriers to the salmon ascending the streams. The present law requires that in event of the construction of such dams that a fish-way be left so that the fish may ascend. . . . Immediate steps should be taken to remedy the matter. (5)

Although it was illegal to block salmon from ascending a river, dam building remained an attractive proposition.

Dams on the Elwha River in Olympic National Park provide the classic case of enduring, illegally constructed dams. Located on the northern end of Washington's Olympic Peninsula, the Elwha harbored all five North American species of Pacific salmon, including legendary monstrous chinook that reached over 100 pounds. The river flows from the interior of the Olympic Mountains through a narrow gorge before dropping to empty into the Strait of Juan de Fuca. The opportunity to dam the river at the gorge to produce power (for markets not yet then in existence) led Thomas Aldwell, a Canadian with backing by Chicago investors, to dam the Elwha. Built between 1910 and 1913, Aldwell's first dam lacked both provision for fish passage and a solid foundation. It failed because of engineering shortcomings but was soon rebuilt, again without the required fish passage.

This illegal fish barrier created a political problem for the newly elected governor of Washington, Ernest Lister. His creative fish commissioner, Leslie Darwin, came to the rescue. Darwin proposed to Aldwell's company that they build a fish hatchery instead of a fishway. Although this would not satisfy the law, Darwin saw a novel way around this technicality. He suggested that if the company built a hatchery that was physically connected to the dam, then the dam could be considered an official, state-sanctioned fish obstruction for the purpose of supplying the hatchery with eggs. Governor Lister liked the idea so much that he persuaded the state legislature to endorse building hatcheries instead of providing for fish passage at new dams.

Completed in 1915, the Elwha hatchery turned out to be a complete failure and was abandoned in 1922. But the precedent had been set, as dams began to multiply and hatcheries became the backbone of state fishery management. Many hatcheries were positioned to block native salmon from moving upstream past the hatchery in order both to help round up returning adults and collect their eggs and to keep fish from getting into and fouling the water supply used by the hatchery.

At the opposite end of the Olympic Peninsula, the city of Tacoma built Cushman Dam in 1926, blocking the North Fork of the Skokomish River. The lack of fish ladders—or even a hatchery under Darwin's novel interpretation of the law—did not impede the project. Virtually the entire upper river was diverted through tunnels to cross beneath a ridge, pass through turbines, and then spill into Hood Canal. Although the flow diversion drained the North Fork above the bypass,

"Too bad the poor fish can't do this!" An illustration from 1936.

the Washington State Department of Fisheries didn't challenge construction of the dam or take legal action against the city of Tacoma.

In other cases, fishing interests mobilized to fight against potential barriers to fish passage. Fisheries agencies were particularly alarmed when a private company, the Washington Irrigation and Development Company, applied for a license to dam the Columbia River at Priest Rapids in January 1924. Noting that at least half the Columbia River salmon spawned upstream of the proposed dam, federal and state fisheries interests submitted a brief opposing the dam to the Federal Power Commission, which was the licensing authority for dam building. Predicting disaster for the salmon and thousands of people dependent on them should the dam be built, they entitled their brief *Save the Columbia River Salmon*. Despite these efforts, less than a decade later the first dam, Rock Island, was built across the Columbia.

The effects of the adverse impacts of water diversions on salmon were well known in the Pacific Northwest. In 1916 a U.S. Bureau of Fisheries agent named Dennis Winn had investigated the effects of irrigation ditches on salmon losses in central Washington's Yakima River. Winn found that during the critical downstream July migration of juvenile salmon, more than 90 percent of the river's flow was diverted into irrigation ditches. He counted an average of twenty fish per acre in the fields. Later, Charles Pollock, the Washington State supervisor of fisheries, concluded in his 1932 annual report to the state legislature, "Practically all established power and irrigation projects have already taken a large toll on fish life" (28).

In 1937, a report of the U.S. commissioner of fisheries identified the effects on the downstream migration of juvenile salmon of water diversions for agriculture as threats to the Columbia River salmon:

> With the rapid development of agriculture and the increase in acreage requiring irrigation, practically all the major streams in eastern Washington have been tapped for irrigation purposes. These streams and their tributaries have in the past supported excellent runs of salmon and steelhead trout, but now only fragments of the former runs remain. It is evident, then, that these unscreened ditches and canals, and similar irrigation systems in Oregon and Idaho, constitute one of the many factors destructive to the fisheries of the Columbia and that steps must be taken to prevent this loss. (1937, 52)

The commissioner went on to state that he saw the problem as one of enforcement, rather than inadequate laws or regulations: "Laws of the States of Washington, Oregon, and Idaho contain provisions for the screening of irrigation canals. . . . For various reasons, economic and otherwise, the provisions of these State laws have not been uniformly enforced in such a manner as to provide for the screening of all or even a substantial part of the existing irrigation diversions" (53–54). Though the laws remain on the books, unscreened irrigation ditches continue to take a toll on migrating salmon throughout the region.

The year after the commissioner's report, biology professor Henry Ward investigated the effects of dams, flow diversions, and hydraulic mining for the Oregon Department of Geology and Mineral Resources. Ward found that dams constructed to produce electricity both impede the upstream migration of adult salmon and endanger juvenile salmon on their way downstream as they pass through turbines. He noted that dams raised the temperature in the sluggish water upstream of dams and that water temperatures in fishways approached the upper limit at which salmon could survive. Ward further noted the effect of unscreened irrigation ditches on migrating juvenile salmon.

> Young fish have been watched often entering such ditches, moving freely down the current, accumulating in deeper holes when the water was cut off, or found dead in irrigated fields. They are seen in miners' settling basins or power-plant reservoirs, are torn to sheds [*sic*] in turbines. . . . It is immaterial whether the diversion ditch serves a power plant, an irrigation project, a mining enterprise or some other purpose, the fish, young and old, which enter it are condemned to destruction. While the number tempted to enter at any particular moment may be small, it must be remembered that such ditches work day and night until shut off and the total count of fish destroyed is unquestionably large. (Ward 1938, 9)

Irrigation diversions on the Rogue River, like those on the Yakima River decades before, lacked screens to keep salmon out of the fields, even though Oregon state law stipulated that diversion ditches be screened to prevent fish from entering. Ward considered unscreened irrigation diversions to represent a needless waste of fish. Enforcement of laws requiring screens on irrigation diversions remains a sensitive issue in the Pacific Northwest.

Although planners thought about the problem of getting salmon upstream to the spawning grounds, and thoughtfully designed salmon ladders seemed to work pretty well (when maintained), the problem of juvenile migration back to the sea remained vexing. Getting fish over a dam seemed like a simple engineering problem, but survival of juvenile salmon during downstream passage through a series of lakes and dams was more complicated. Though solutions were elusive, the problem was recognized as critical to the survival of healthy salmon runs. Still, dam operators on the Columbia neglected to make allowances for getting juvenile fish down the river in the time that they are physiologically adapted to make the journey.

Juvenile salmon evolved to ride downstream on the Columbia's swift current during spring floods. They are genetically programmed to mature so that when they get to the bottom of the river system, they are ready for the physiological changes that allow them to survive the transition to salt water. If they are forced to swim through lakes rather than roll along on a mighty current, the going is too slow. If the trip takes too long, they mature too fast, and many die before they reach the sea.

Society's gradual decision in effect to run the Columbia River as a giant power generator involved changing the timing of peak flows and otherwise profoundly altering the seasonal rhythm of the river. Peak runoff does not coincide with peak electrical loads in the Pacific Northwest. So power generators on the Columbia store water behind dams during spring high flows and release water during periods of high demand for electricity. Dam operations changed the Columbia from a river with high spring flows that helped speed juvenile salmon down to the ocean into a long series of lakes that slowed this crucial migration. For all the expenditure and planning for allowing salmon to make it over or around dams on their way upriver, operating dams so as to facilitate the equally crucial downstream migration remained a low priority.

Turbines also kill salmon. An estimated 10 to 15 percent of downstream-migrating juvenile salmon that go through the turbines of power-generating dams are killed at each dam. For stocks from rivers low on the Columbia system, and which only pass through one or two dams, this is not such a big deal. But for stocks from the upper Columbia River basin it becomes a very big deal. With just a 10 percent loss at each dam, a population that ran through ten dams would lose two thirds of the original run; 15 percent mortality at each dam would result in only one out of five fish making it past ten dams.

In 1927, Congress directed the Army's Corps of Engineers to study the development potential of the Columbia River basin. Four years later, the corps reported that there were ten promising locations for dams to provide for water power, flood control, and irrigation. The following year, Franklin D. Roosevelt, who was running for the presidency, promised a cheering crowd in Portland, Oregon, that the next hydroelectric development supported by the federal government would be on the Columbia River. Any concern over the potential impact on the salmon paled against the Depression era promise of a revitalized regional economy.

In 1933 construction started on the Bonneville and Grand Coulee dams, projects that were viewed as emergency public works whose primary purpose was providing jobs and stimulating a stagnant economy. Bonneville Dam was intended to improve navigation on the Columbia, and so it was to be operated by the Army Corps of Engineers. Grand Coulee Dam was built to provide irrigation for arid eastern Washington and was to be operated by the Bureau of Reclamation. Hydropower was a secondary reason for building both dams.

Some attempts were made to mitigate fisheries concerns during development of the Columbia River. Bonneville Dam, located low on the Columbia, was built with a fish ladder. In contrast, Grand Coulee Dam, located in the upper reaches of the basin, is an utterly impassable 550-foot high structure that would dwarf Egypt's Great Pyramid. The plan for solving the fish problem for Grand Coulee was to catch the salmon from above the dam and release them into tributaries below the dam. The fish were to adapt to these more conveniently located rivers.

Before construction of Grand Coulee Dam, about 5 percent of the Columbia River salmon spawned upstream of the proposed dam site. Fish from these runs were experimentally transplanted to below the dam site in the Okanogan, Methow, Entiat, and Wenatchee rivers. At first the experiment appeared successful, and fisheries managers were lulled into optimistically thinking that transplantation could be relied on throughout the Columbia River basin. Later attempts at transplantation were less successful, and the number of new sites to transplant salmon into kept declining as the basin developed.

The initial plan for developing the Columbia River did not include provisions for fish passage. Nonetheless, the Army Corps rejected the idea of a high dam at The Dalles, on the lower Columbia River because such a dam would block and thus destroy salmon runs on virtually the entire Columbia River system. In 1936, the Corps of Engineers recommended postponing proposals to dam the lower Snake River, arguing that such efforts should wait until the effects of the Columbia River dams on the salmon had been ascertained. Two years later, however, in an about-face orchestrated by shipping and agricultural interests, the Corps released a new report that advocated damming the Snake River. Now focused on turning swift rapids into easily navigable lakes to aid shipping between Lewiston, Idaho, and the Pacific Ocean, the Corps presented a proposal to build a series of dams connected by locks.

The 1937 Bonneville Project Act created the Bonneville Power Administration (BPA) to market power from the new dams. The act directed the BPA to encourage power use and give preference in power sales to public utilities. The mighty Columbia presented an enormous source of power and could produce far more energy than the region demanded. So the BPA kept rates low to encourage power consumption. Lacking customers, they built a market by constructing transmission lines so that rural areas were electrified within a decade, which dramatically improved the quality of life for people throughout the region. As part of its publicity campaign it hired a young folksinger, Woodie Guthrie, to compose songs extolling the dams and the power they would bring. Meanwhile, the U.S. Bureau of Reclamation developed plans to build dams all along the Columbia, Snake, and Willamette rivers.

From the start there was concern over the potential effect of hydroprojects on Columbia River salmon. In commenting on the proposed Bonneville Dam, the Stanford University biology professor Willis Rich pointed out that those proposing the dams had not determined whether provisions for the passage of fish past the dam would be successful. He expressed concern as to whether salmon would actually scale the fish ladders designed for the dams. But doubts about the potential effects on salmon were waved aside with assurances that fry from hatcheries would more than make up for any loss.

In 1933, the year that the first Columbia River dam was completed, a real estate developer, Frank T. Bell, was appointed United States commissioner of fisheries. Prior to his appointment, Bell promoted development of the Columbia River basin and in particular the construction of Grand Coulee Dam. Bell was an effective mouthpiece for development interests, who even as the dams were being constructed wrote, "[W]e have no reason to believe that the Columbia River salmon are in danger of extinction. We feel confident that the preservation of the great national resource of Columbia River salmon is assured" (Bell 1937, 46). Was Bell's overconfidence due to misplaced faith in fish ladders? Or hatcheries? Or was he just blowing smoke?

Concern over the potential impacts of plans to develop the Columbia River basin led Congress to direct Commissioner Bell to assess the effect of Bonneville Dam on salmon runs. His official report to the U.S. Senate Committee on Commerce acknowledged that the dam would present "a barrier which, if not surmounted, will destroy the major portion of the fish supply . . . [and] affect perhaps 75 percent of the total salmon supply of the region" (U.S. Commissioner of Fisheries 1937, 2). The commissioner allowed that there was substantial uncertainty in how the delay in traveling upstream due to passing the dams might affect salmon runs, stating, "While this delay is probably not serious when only one or two dams are involved, no one can say at the present time what may happen if the salmon are required to surmount nine dams" (U.S. Commissioner of Fisheries 1937, 51).

Commissioner Bell recommended increasing support for research on the natural history and ecological requirements of Columbia River salmon because

> faced with the emergency presented by the construction of Bonneville and Grand Coulee Dams, fundamental information has been found to be so incomplete that vast programs of development were of necessity undertaken without assurance of their possible effects on the fish supply. . . . As a result, aside from blind restriction of commercial activity, the protection of individual runs menaced by virtual extinction must at the present time be left to chance. (75–76)

This is a striking admission on the part of the commissioner of fisheries that he was powerless or unwilling to delay the Corps' program of dam building or was uninterested in trying to do so. The plan to protect Columbia River salmon was to trust in chance.

During the Depression, the need for power and jobs trumped uncertainty over potential effects of dams on salmon. Even so, in 1938, a year after receiving the commissioner's report, Congress passed the Mitchell Act, authorizing and funding measures to protect Columbia River salmon while the Columbia River was developed. The Mitchell Act gave the federal government authority to construct salmon hatcheries, funding of research to facilitate conservation of the fishery, construct and maintain habitat improvements, and "all other activities necessary for the conservation of fish in the Columbia River Basin" In granting such broad authority, Congress was not authorizing the extinction of the Columbia River salmon.

Yet despite the Mitchell Act, the agencies operating the dams that followed did not adjust operations to facilitate salmon migration. Although no one made an explicit decision to sacrifice salmon to power development, operation of the system ensured that outcome. Routinely adopting optimistic assumptions about the potential effects of projects and operations on salmon, and invoking the need for more study before making changes they did not desire, agencies running the dams sacrificed the Columbia River salmon by giving top priority to navigation, irrigation, and hydropower.

This implicit decision did pay some dividends. Whatever else one may think of the impact of the Columbia River dams, they helped win the Second World War, which was fought as much in factories as on battlefields in Europe, Africa, and Asia. As bombs reduced European cities to rubble, unscathed factories in the United States increased production until by the war's end the logistical balance was profoundly lopsided in favor of the Allies. Nowhere was this imbalance greater than in the production of the airplanes that destroyed the capacity of the Axis nations to wage war. The BPA played a decisive role in winning the war of production and in transforming the United States from an agricultural society into a global superpower.

In December 1939, just three months after Hitler invaded Poland, the BPA signed a contract to supply cheap public power for an ALCOA (Aluminum Company of America) plant in Vancouver, Washington. With Hitler overrunning Europe, Congress approved President Roosevelt's request to dramatically increase U.S. aircraft production to a staggering 50,000 planes per year. The 5-billion-dollar Defense Appropriations Act of June 1940 led to half a billion dollars in defense contracts going to the Pacific Northwest. By June 1943, war production was using more than 90 percent of the power generated by BPA's dams. Now there was a market for BPA power. The tamed Columbia River fed aluminum plants that supplied shipyards and aircraft plants building naval and aerial armadas. The war also fueled efforts to construct additional dams on the Columbia.

Fisheries agencies supported building the Dalles Dam because inundation of Celilo Falls was seen as a way to end Indian fishing. Samuel J. Hutchinson, the regional director of the U.S. Fish and Wildlife Service, considered elimination of the Indian fishery at Celilo Falls a good reason to build the dam at The Dalles because "the beneficial effects [of the dam] would compensate for the detrimental conditions that exist there at present. In brief, it would be easier for the fish to go over a ladder in the dam than to fight their way over Celilo Falls. The Indian commercial fishery would be eliminated and more fish would reach the spawning grounds in better condition" (Cone and Ridlington 1996, 206).

A 1958 U.S. Fish and Wildlife Service report on the Columbia River fishery development program called "the inundation of Celilo Falls in the Columbia River near The Dalles, Oregon . . . a memorable event. Removal of this partial barrier by backwater from The Dalles Dam eliminated forever an intensive historic Indian fishery" (1958, p. 3). In the same report, however, the Fish and Wildlife Service acknowledged that though they viewed the Indian fishery as a threat needing to be eliminated, they wanted to foster the growing sport fishery through supplementation and management of the fishery.

A decade after passage of the Mitchell Act, the U.S. Fish and Wildlife Service and the three states of the Columbia River basin—Oregon, Washington, and Idaho—agreed to a program of fishery enhancement measures. In its 1948 report recommending construction of the Dalles and John Day dams, the Army Corps of Engineers concluded that an aggressive fishery-development plan on the lower Columbia River was needed to maintain commercial salmon runs. Most of the measures proposed had a familiar ring to them: removal of barriers to fish passage, construction of fishways, pollution abatement, and more hatcheries. But two new elements were also agreed upon: Runs at risk would be transplanted to new locations, and fish refuges would be established where salmon preservation would be paramount. A backup plan to ensure survival of salmon seemed within reach on the Columbia.

The plan for developing the upper Columbia River basin called for maintaining the Cowlitz River in Washington State as a salmon sanctuary. But Tacoma City Light was eyeing the Cowlitz as a way to meet its own growing demand for electrical power. In late December 1948, the city of Tacoma applied to the Federal Power Commission for licenses for two dams on the river. The utility proposed a novel solution for getting salmon over the downstream dam, which was to be twice as tall as any dam with a successful fishway ever built. The fish were to start climbing over the dam on a fish ladder. At the top of the ladder, only partway up the dam, they would be trapped and transferred via a tramway to trucks that would haul them the rest of the way over the dam. Unconvinced that this approach would actually work, state fish and game agencies blocked the license needed to build the project. In 1949, the State of Washington passed a law prohibiting dams taller than 25 feet on all but two tributaries of the lower Columbia River. Undeterred, the Federal Power Commission dismissed these concerns and issued licenses for both Cowlitz river dams in November 1951, confident that with ingenuity and effort, the salmon problem could be solved.

But that is more easily said—or even mandated—than done. For no one branch of the regulatory system can get a handle on the problem. Salmon cross all of our jurisdictional boundaries, meaning that everybody controls a piece of the salmon puzzle but no one is in charge of putting it together. Swimming up through estuaries and big rivers into smaller streams—sometimes literally into somebody's backyard—their welfare is impacted by the decisions of all levels of government. The decisions of federal agencies, states, counties, city councils, and the actions of individual property owners all influence the productivity of salmon habitat.

The overlapping authorities across a dizzying array of jurisdictions are rarely coordinated and often act at cross purposes such that even broad laws passed by Congress that specifically mandate the protection of salmon are undermined or thwarted by citizens or other governmental agencies. The problem of a complex and confusing overlap of jurisdictions fostered uneven and poor enforcement of laws and federal regulations that contributed to the gradual decline of salmon across the Pacific Northwest.

It certainly didn't help the salmon that they were a low priority for the agencies running the Columbia River. The 1945 Rivers and Harbors Act, which adopted the Army Corps of Engineers' plan for the Snake River, required that the project allow salmon "free access to their natural spawning grounds." As in the Mitchell Act, such wording shows that Congress did not intend development of water projects in the Columbia River basin to exterminate salmon runs. While developing the river, the Army Corps of Engineers repeatedly expressed its intent to conserve salmon "to the maximum practicable extent," to have its operations result in "minimum interference" with salmon habitat, and to incorporate into its projects "the best possible means for passing salmon upstream and downstream." However well intended, such considerations were always an afterthought, resulting in minor adjustments to plans tailored to achieve higher priorities.

Alvin Anderson of the Pacific Marine Fisheries Commission, writing in 1950, described in "Shall We Have Salmon, or Dams, or Both?" the imbalance between the influence of fishery interests and government agencies promoting development of the Columbia River dams.

> Yes, it is possible to have salmon and dams when the lobby and legislative actions for the salmon are as strong as the lobby and legislative actions for the interests destroying salmon. . . .
>
> The practice of turning down a specific project because it would mean the loss of the salmon resource has carried such vicious implications to political representatives as to force their acquiescence, and even their pressure, for the instigation of disastrous projects.
>
> Dam construction consistently indicates the expenditure of large sums of money which turn the wheels of the cement, steel, transportation and other industries. . . . Strength of representation from these demands finds fishery representation a weak opponent. (449–51)

In the early 1950s, fearing a rerun of New England's salmon debacle, regional fish and game agencies delivered a joint statement to the Northwest Governors' Power Policy Committee in which they argued for a cautious approach to dam building: "Within the last century salmon runs in the New England states were unhesitatingly sacrificed for the purpose of industrial development. In many cases the salmon would now have greater value, if they existed, than the industries which replaced them" (Netboy 1958, 101).

But commercial interests wanting to transform Lewiston, Idaho, into a major shipping port were pushing efforts to turn the wild Snake River, the largest tributary of the Columbia, into a series of placid reaches that barges could navigate. In 1945 Congress authorized construction of Ice Harbor Dam, the cornerstone of the Corps of Engineers' plan for a series of dams along the river. The battle over funding raged for years, and at first, fishery interests managed to block allocation of funding to construct the dam.

Testimony before the House Appropriations Committee captures the debate over the proposal to include funding to begin construction of Ice Harbor Dam in the federal budget for fiscal year 1953. Representative Louis C. Rabaut of Michigan interrogated Brigadier General C. H. Chorpening, the chief of the Army Corps of Engineers: "The Congress has disapproved this project so many times that it has gotten to be a habit. What has happened between last fall and now that causes you to request this item again?" (U.S. House of Representatives, 1952, Part I, 177)

Chorpening replied that nothing had changed, and then proceeded to read a classified letter off the record. The key difference was that this time around the proposal invoked the need to supply power to the Hanford works of the Atomic Energy Commission. Representative Homer D. Angell of Oregon, a major proponent of the dam, argued that delay in constructing Ice Harbor Dam would handicap national defense (Part I, 270).

General Chorpening testified that the benefit-to-cost ratio for the proposed dam was less than 1.2 to 1, an unimpressive estimate that did not include any accounting for lost salmon. In addition, without the navigation benefit from the dam, which could only accrue if all four of the other proposed Snake Rivers dams were built, the cost of Ice Harbor Dam would equal its projected benefit. After presenting the Corps's analysis, Rabaut asked Chorpening about the annual value of the Columbia River salmon catch, which at the previous year's hearing the Corps had testified was worth $8 million—more than the value of the annual benefits expected from the dam. In response, Chorpening steadfastly maintained that "there will be no difficulty at this project in the proper handling of the fish problem" (U.S. House of Representatives, Part I, 178), although he acknowledged that "there is the killing of the little fry going downstream; we have not been able to solve that" (Netboy 1958, 76). Once again, the plan was to figure something out later.

William Hagen, the man in charge of salmon propagation for the U.S. Department of Commerce's fisheries development program on the Lower Columbia River, testified that 15 percent of the downstream migrating juvenile salmon were killed at Bonneville Dam. The director of the Oregon Game Commission, P. W. Schneider, the chairman of the Fish Commission of Oregon, John C. Veatch, and Oregon's director of fisheries, Arnie J. Suomela, all testified that a 15 percent mortality on downstream migrating juvenile salmon at each of the four proposed Snake River dams would wipe out three quarters of the Snake River salmon. This would seriously alter the Corps's benefit-to-cost ratio.

Schneider summarized the position of the fishery interests. They did not oppose the development of water resources in the Columbia River basin but felt that the dams that did not endanger salmon runs should be built first, if only to buy time to better determine how to solve the fish problem. Schneider said, "We sincerely fear that the construction of these lower Snake dams at this time would doom for all practical purposes a major portion of the Columbia runs" (U.S. House of Representatives, 1952, Part 2, 33).

Veatch and Suomela pointed out that the Columbia and Snake river salmon were well worth protecting, pleading, "The Nation simply must not repeat the mistakes that have needlessly destroyed some of its resources" (U.S. House of Representatives 1952, Part 2, 36).

Once again, the House Appropriations Committee rejected funds for construction of Ice Harbor Dam. The chairman of the House Committee on Appropriations, Representative Clarence A. Cannon of Missouri, offered an eloquent explanation of the majority opinion:

> The construction of this dam means eventually the complete extinction of a species of salmon which thereafter can never be resuscitated or recreated. Only God in His infinite power and wisdom can create a new species of animal life, and when that is once destroyed there is no power on earth that will reproduce it. . . . The total extinction of life for all time to come is something we cannot even contemplate, regardless of the need for power for the few years that will be required to develop some method of avoiding this permanent restriction of an important food supply. (U.S. House of Representatives, 1952, Part 2, 879)

Support for federal power projects in the Northwest stalled until Democrats regained control of Congress in 1955. By clever parliamentary maneuvering, Northwest Democrats avoided further hearings and out-maneuvered Republican opposition to the dams. Senators Warren G. Magnuson of Washington and Wayne L. Morse of Oregon convinced a joint Senate-House conference committee to slip $1 million for Ice Harbor into the 1956 Army Civil Works Appropriation bill. The first domino on the Snake River fell, and dams started marching up the river.

Because loss of fish could not be predicted accurately in advance, the fears of fisheries interests always took a backseat to development interests. You *could* predict (and bank on) the value of increased power or irrigation or shipping. Given that fishing interests could not demonstrate in advance how any one dam, let alone a whole series of dams, would affect the number of salmon, their concerns were discounted. Such problems create policy biases in which risky actions continue until uncertainty is eliminated—something that can be quite difficult to do even with well-organized and -funded studies.

The last of the Army Corps of Engineers' dams on the Snake River, Lower Granite Dam, came on line in 1975. When a severe drought in 1977 threatened disaster for salmon runs in the upper Snake River

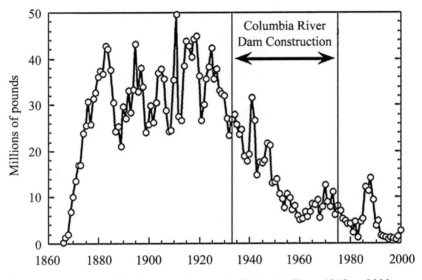

Graph showing annual salmon catches on the Columbia River, 1860 to 2000.

basin, political pressure forced the reluctant Army Corps and the BPA to release flows from dams to support downstream migration of young salmon. The agencies maintained, however, that they had no authority to spill water over the dams for anything other than power production.

The following year the National Marine Fisheries Service started to evaluate whether to list certain runs of Columbia River salmon under the Endangered Species Act.

As conflict grew between those demanding more electrical power and those interested in protecting dwindling salmon runs, Congress passed the Northwest Power Act in 1980. The act directed federal agencies operating the Columbia River power system to "treat fish and wildlife as a co-equal partner . . . on a par with power needs." Under the 1980 act, the Northwest Power Planning Council (NPPC) was created to implement an interstate program to protect and enhance Columbia River salmon runs, and other wildlife. Congress mandated that the NPPC use the "best available scientific knowledge" and favor biological interests over economic ones. Congress also mandated that the NPPC improve salmon survival through increased flow releases capable of meeting "sound biological objectives." The idea of giving salmon equal, let alone a higher priority than power production promised to revolutionize operation of the Columbia River hydropower system. In the wake of the Northwest Power Act, the National Marine Fisheries Service optimistically suspended their initial review of possible Endangered Species Act listings for Columbia River salmon.

The lofty rhetoric and promises of fish-friendly operations proved hollow. The Corps of Engineers refused to implement NPPC recommendations for salmon passage and dam operators consistently failed to provide water flows needed for salmon conservation as recommended by agencies and tribes. Instead, dam operators treated reservoir refilling as a higher priority than salmon and reduced the amount of water dedicated to aiding downstream migration of Snake River salmon. Instead, juvenile salmon were to be trucked and barged around the dams.

How was that possible? The Northwest Power Act required all interests to give the protection of salmon highest priority. How could a law passed by Congress become impotent so quickly? Perhaps through the creation of an administrative system with the power to decide that driving salmon around dams constitutes not only a scientifically credible way to accommodate the needs of migrating salmon but employs the "best available science."

By the early 1990s the Columbia River runs had dwindled from 11 million to 16 million fish prior to European settlement to about 2 million fish supplemented by over 100 million hatchery-raised fry released into the river each year. With wild runs continuing to decline, the National Marine Fisheries Service was petitioned to list the three Snake River stocks and one Columbia River stock under the Endangered Species Act. In 1991, just four sockeye salmon, three males and a female, returned to spawn at Redfish Lake. Finally convinced that the situation was dire, the federal government listed the Snake River sockeye as endangered. They acted just in time to welcome Lonesome Larry back to Redfish Lake in 1992. Listing of chinook salmon soon followed.

Endangered Species Act listings of salmon in the Columbia River basin led the Northwest Power Planning Council to adopt a new strategy for managing the fisheries in December 1992. State and Tribal fisheries managers recommended increasing flows past dams to assist juvenile salmon migration. The new strategy adopted by the NPPC recommended studying the potential to draw down reservoir levels on the Snake River dams to facilitate salmon migration. Implementation of revised flow management would begin in 1995 if the measures proved economically feasible. Power interests of course challenged the benefit of increased flows, arguing that the uncertainty of the benefits mandated maintenance of the status quo.

In the meantime the National Marine Fisheries Service certified that dam operations represented no threat to the Snake River salmon, even though the agency acknowledged that 70 to 90 percent of juvenile salmon would die trying to get past the dams. Although this no-jeopardy opinion was rejected as arbitrary and capricious by a federal judge, the ESA listings had little effect on the operations of the Columbia and Snake River dams. In rejecting the NMFS opinion, District Judge Malcolm Marsh, in deciding a suit brought by the Idaho Department of Fish and Game against the National Marine Fisheries Service, declared, "Instead of looking for what can be done to protect the species from jeopardy, NMFS and the action agencies have narrowly focused their attention on what the establishment is capable of handling with minimal disruption" (1994, 900).

Environmental groups under the umbrella of the Northwest Resource Information Center sued the Northwest Power Planning Council over its salmon strategy, and in 1994 a three-judge panel of the Ninth Circuit Court of Appeals ruled that the NPPC had violated the Northwest Power Act by ignoring recommendations from fisheries, agencies, and Indian tribes. In their ruling they stated, "The Council's approach seems largely to have been from the premise that only small steps are possible, in light of entrenched river user claims of economic hardship. Rather than asserting its role as a regional leader, the Council has assumed the role of a consensus builder, sometimes sacrificing the Act's fish and wildlife goals for what is, in essence, the lowest common denominator acceptable to power interests" (*Northwest Resource Information Center v. Northwest Power Planning Council* 1994, 1395).

Even with longstanding concern over protecting salmon runs, and a chorus of warnings about the effects of dams on fish, only near the end of the period of dam building, in the 1960s, did concern over salmon actually stop plans for a dam. In the late 1950s, power companies proposed building High Mountain Sheep Dam, just below the Snake River's confluence with the Salmon River. The Department of the Interior contested the application for a license because the dam would destroy the salmon runs of the Salmon and Clearwater rivers. Nonetheless, the Federal Power Commission approved the proposal in 1964. Secretary of the Interior Stewart Udall directed the Department of the Interior to take the unusual step of suing the Federal Power Commission to block the dam.

Such open warfare among federal agencies was unprecedented. But by the time the case reached the U.S. Supreme Court, Udall's concerns seemed justified: three dams built by the Idaho Power Company had already extinguished salmon runs on the upper Snake River. In a decision written by Justice William O. Douglas, the Court sided with Udall, noting, "The importance of salmon . . . in our outdoor life as well as in commerce is so great that there certainly comes a time when their destruction might necessitate a halt in so-called 'improvement' or 'development' of waterways. The destruction of anadromous fish in our western waters is so notorious that we cannot believe that Congress . . . authorized their ultimate demise" (*Udall v. Federal Power Commission* 1967, 437–438).

This decision put a halt to construction of High Mountain Sheep dam, but it did not end the fight over the lower Snake River dams; three more were eventually built by the Army Corps of Engineers between 1969 and 1975. The lower Snake River dams brought the number of major dams built on the Columbia River system between 1933 and 1975 to eighteen, an average of one every other year.

In its original plan for the Snake River, the Army Corps of Engineers did not plan for dams to provide year-round navigation. They anticipated a two-month period when navigation for most vessels would be blocked by ice or floods. This period is very similar to what is being sought today to facilitate downstream migration of juvenile salmon on the Snake River system. But the original plan is now viewed as an unacceptable economic hardship. Convinced that water spilled is water wasted, dam operators consistently worked on the assumption that salmon losses would be offset by hatchery production and therefore that inconvenient changes in dam operations were not really necessary anyway.

In 1995, in response to the ESA listing of Snake River salmon stocks, the National Marine Fisheries Service (NMFS) asked the Army Corps of Engineers to study the possibility of removing the lower Snake River dams. In the Army Corps report completed after four years of studying the problem, the U.S. Fish and Wildlife Service concluded that dam removal would best provide the highest certainty of saving the fish. Scientists from other agencies, including NMFS, also recommended dam removal. Oregon's Governor John Kitzhaber astounded the region's political establishment by calling for breaching the four Snake River dams. In May 2000, an internal draft of the NMFS plan instructed the Army Corps to begin planning to remove four lower Snake River dams. But by December 2000, when the NMFS issued its final plan, the focus had changed to delaying a decision on dam removal for at least eight years and instead relying on pilot projects, voluntary habitat improvements to be done by state, tribal, and private groups, and transporting fish around dams in trucks and barges.

Environmental and fishing groups sued the federal agencies in 2001, alleging that the NMFS plan relied on unspecified future actions to reduce impacts of habitat, hatcheries, and harvest, but ignored the impacts of the dams themselves, which even the NMFS plan acknowledged jeopardized survival of salmon stocks listed under the Endangered Species Act. In a May 7, 2003, opinion rendered in Portland, U.S. District Judge James Redden ruled that the NMFS plan failed to adequately protect endangered salmon runs because the actions relied upon in the plan were "not reasonably certain to occur." Advocates of dam removal tried to spin the decision as backing their agenda. Oregon Senator Gordon Smith fired back in a May 15 press release vowing to do everything in his power to block moves to breach the dams. Judge Redden is reported to have said, "I have a recurring nightmare that we're still talking up here about the fish problem and somebody catches the last one."

In the summer of 2001 there was a series of power outtages in California, and water intended to supplement flows to help downstream salmon migration was instead spilled during peak demand periods to sell power to California at astronomical prices. Yet this was a low-water year when the extra flows were critically needed by the salmon. Even after Endangered Species Act listings, the downstream migration of juvenile salmon remains a lower priority with those running the Columbia River power grid than does cashing in on power deregulation.

Columbia River salmon were sold down the river for the promise of cheap electricity. Fulfilling this promise, the BPA dams on the Columbia River system transformed the economy of the Pacific Northwest, helped win the Second World War, and turned the dry interior of the region into some of the most productive farmland on the planet. Yet today the region has skyrocketing power rates, and both salmon and power are expensive. To be sure, the taming of the mighty Columbia River catalyzed regional development, albeit at the price of now legendary salmon runs. Similarly, twentieth-century forest clearing, rural development, and urbanization led to equally deadly changes to many other salmon rivers throughout the Pacific Northwest.

chapter 10

RIVERS OF CHANGE

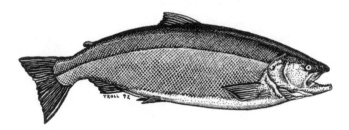

The maintenance of conditions favorable to fish life in our mountain streams is largely dependent upon the preservation of the forests.

S. B. Locke, *U.S. Bureau of Fisheries Document* 1062, 1929

THERE IS SOMETHING TIMELESS ABOUT A RIVER. UNLESS the view includes something really strange like burning water or a big dam, it is easy to imagine that a river has always been pretty much like what we see today. The seductive allure of flowing water fosters an assumption that rivers maintain their character over time as they carry away water, sediment eroded from hillsides, fields, and mountains, and whatever else we dump into them. But rivers are dynamic systems. Depending on what enters its channel, a river may serve as a sewer or a life-giving artery.

As rivers adjust to changes in their load of water, sediment, and wood, the resulting changes in habitat type, abundance, or quality can dramatically alter the health, number, and kinds of river-dwelling organisms. A legacy of prehistoric change characterizes the environmental history of Puget Sound rivers, but these natural disturbances since the last glaciation were either small, relatively short-lived, or confined to only some river basins. With generations of reserves at sea in any given year, salmon thrived in the dynamic, ever-changing landscape even as local populations occasionally crashed or temporarily disappeared.

Pacific Northwest rivers today bear little resemblance to those that sustained salmon-based economies and cultures of native peoples centuries ago. After several generations both individual and collective memories of the land get pretty murky, and the memory of the landscape among those still alive extends back little more than half a century. We think we know the world we see today, vaguely remember the world of our childhood, and can sort of imagine our grandparents' world. But few people are aware of the extent of historical changes to the Pacific Northwest's rivers. Fewer still have any real sense for what rivers were like before the signing of the treaties that unleashed the changes still sweeping across the region.

It is quite difficult to find places to go to get a feeling for what the rivers of the Pacific Northwest were like before European settlement. Even rivers that have benefited from restoration efforts remain tamed and sanitized versions of their former selves. A canoe trip on the lower Nisqually River, which flows into southern Puget Sound just north of Olympia and which serves as the border between the Nisqually Indian Reservation and the U.S. Army's Fort Lewis, is as close to a trip down the rivers of Lewis and Clark's day as can be found around Puget Sound.

The 1854 Treaty of Medicine Creek reserved the lower reaches of the river for the Nisqually Tribe. Consequently, while streamside forests on other Puget Sound rivers were cleared by the early 1900s, the valley bottom along this stretch of the Nisqually remained relatively undisturbed. The U.S. Army annexed the north side of the valley in

1917, and still uses it as a buffer for its artillery practice range. The lower Nisqually River remains virtually bypassed by history, left close to its natural state except for occasional heart-stopping artillery barrages.

In the summer of 1998, Brian Collins, Tim Abbe, and I canoed the lower Nisqually to study the influence of logjams on the river. Floating down this river is a real treat for us logjam connoisseurs. Grand old cottonwoods growing on the floodplain are big enough to remain stable when, once undermined by bank erosion, they fall into the river. The way logs stack against each other reveals that wood moving through the channel racks up on stable logs as though woven into a natural tapestry.

Logjams capture more logs and further obstruct flow, feeding a runaway process that leads to the immense piles of wood that greet a canoe around almost every bend. On our way down the river we found huge jams plugging the entrance to side channels that once served as the main channel. We even found a curious place where aerial photographs show that for a while the river flowed backward for a few hundred yards upstream of where it was completely blocked by a logjam. We counted two thousand logs per mile of channel, more than 90 percent of them racked up in gigantic jams.

Our trip showed us that, much like that of the Queets River in Olympic National Park, the valley bottom along the lower Nisqually River hosts a complex web of channels. In the main channel as well as the side channels, the deepest pools lie under and around the biggest logjams. Although potentially dangerous to an unlucky boater, the logjams on the Nisqually create ideal salmon habitat with lots of deep pools in the main channel, a network of tranquil side channels that flow year round, and an abundance of overhanging and submerged logs to provide cover from predators. As we left the river at the end of the day, fishermen standing shoulder to shoulder along the banks testified to the quality of the river as salmon habitat.

Early accounts of Pacific Northwest rivers paint a portrait unfamiliar to most people, but recognizable by anyone who has floated down the Nisqually or tried to march up wood-choked rivers into the heart of Olympic National Park. In his journal of a Hudson's Bay Company trapping expedition on the central Oregon coast in the summer of 1826,

Alexander McLeod, the expedition's chief trader, described logjams on Oregon's Siuslaw River:

> Jeaudoin and his companion went in Course of the Afternoon some distance up the North Branch of the River but finding the Navigation much impeded by fallen trees they returned at dusk conceiving the Obstacles insurmountable. . . .
>
> [L]ate last night the four men sent up the South Branch returned not being able to proceed owing to the quantity of fallen Trees which obstructs its navigation. (Davies 1961, 167, 171)

Fur trappers like McLeod pursued beaver all over North America, yet coastal Oregon's logjam-filled rivers stymied them. The region's huge trees simply were too big for streams to move. Vast numbers of snags and logjams clogged rivers throughout the Pacific Northwest. Charles Wilkes, commander of the 1838–42 U.S. Exploring Expedition to the Pacific, reported that even on the mighty Columbia, "channels in this river will be always more or less subject to change, from the impediments the large trees drifting down cause, when they ground on the shoals" (1844, 152). To those who take modern wood-poor rivers as a frame of reference, the amount of wood in the prehistoric rivers of the Pacific Northwest is almost unimaginable.

Brian Collins investigates the extent and style of historical changes to Puget Sound rivers in order to re-create as accurate a picture as possible of their original condition. Acting as an environmental forensic detective, he pieces together fragments of information to reconstruct a picture of pre-European contact river conditions. His work takes him to library archives and on canoe trips to survey modern rivers flowing through mature forest, agricultural valleys, and rapidly urbanizing lands.

In the course of his work, Brian scoured archival records of the Army Corps of Engineers to tally the number of logs pulled from Puget Sound rivers in the early days of Euro-American settlement. The Army engineers started clearing Puget Sound rivers in 1880. Though these efforts continue to this day, they peaked between 1890 and 1910, when an average of three thousand snags were cleared every year from the lower Skagit River. Many of these were massive trees; the largest snags pulled annually from Puget Sound rivers were 12 to more than 15 feet in diameter. Over the course of a century 150,000 snags were cleared from just five rivers.

Puget Sound rivers were not unique in their prehistoric load of large wood. Jim Sedell, a researcher with the U.S. Forest Service, pioneered the use of historical records in reconstructing changes in river conditions on the Willamette River, Oregon's second largest river. Sedell found that over 5,000 trees 5 to 9 feet in diameter were cleared from a single 50-mile-long reach in the first decade of the Army Engineers' program to improve the river, 1870–80. All together, more than 65,000 snags and streamside trees were removed from the Willamette from 1870 to 1950—an average of 880 per mile of river. That's about one every six feet.

The 1875 annual report of the Secretary of War (U.S. House of Representatives 1875) described the Willamette as littered with logs "too numerous to count" and described how "[e]ach year new channels are opened, old ones closed; new chutes cut, old ones obstructed by masses of drift [logs]; sloughs become the main bed, while the latter assume the characteristics of the former; extensive rafts are piled up by one freshet [high flow] only to be displaced by a succeeding one; the formation of islands and bars is in constant progress" (766). The Army Engineers noted how logjams not only split the river into four or five main channels but also blocked navigation.

Before the clearing of old-growth forests made possible convenient overland transportation, rivers served as transportation corridors along which communities were organized and lives were oriented. But travel was slow. Snags and logjams made most rivers nearly impassable. Native settlements were located along navigable waterways, including the lower reaches of major rivers. Few people lived inland or up the river valleys. As the regional population exploded, settlers wanted rivers opened to navigation to develop the agricultural potential and forest resources of the fertile valleys and rugged uplands. Spreading inland along river corridors, settlers cleared valley bottoms first as they built a regional economy.

Assigned the task of pulling snags from Puget Sound rivers, the Army Engineers recorded their progress in annual reports to Congress.

In their report for 1875, Major Nathaniel Micheler described the Sno-homish River as "very shallow and filled with snags." Similarly, Assistant Engineer Robert Habersham reported that in August 1879 the Stil-laguamish River was "much obstructed by snags and trees embedded in the bottom, and at six points completely closed by rafts [of logs] which have diverted the current so as to cut out minor channels, forming small islands" (U.S. House of Representatives 1881a, 2610).

The Army Engineers' objective was an unobstructed, single-channel river open to commerce. The biggest logjams formed great rafts of wood that could block even the Skagit River, the largest river draining into Puget Sound. At the site of the modern town of Mount Vernon, a stable raft of logs blocked the Skagit for more than a century before set-tlers cleared it in the 1870s. The jam wasn't just a bunch of logs stuck together in the river. Tier upon tier of logs up to eight feet in diameter, and packed solidly enough to be crossed almost anywhere, formed a stable obstacle that supported a forest of 2-to-3-foot-diameter trees growing on its surface. The forested log raft forced the main channel of the river to flow beneath a natural wooden bridge.

Beneath the raft, dangerous rapids raced between deep sheltered pools described by settlers as full of fish. Spanning the main channel, the Mount Vernon logjam obstructed high flows and created a backwa-ter that annually flooded about 150 square miles of the upstream valley. Perennially submerged wetlands and sloughs provided ideal summer rearing habitat and slow-water refuges for salmon during winter floods.

Valley bottoms began to change as settlers and special "snag boats" designed for the task cleaned logs and logjams from rivers. In 1877, after failing to obtain an appropriation from Congress to remove it, set-tlers started to dismantle the massive Skagit River jam themselves by cutting up the huge logs holding it together. Within a few years the ob-struction unraveled and the extensive network of freshwater wetlands along the Skagit River began to dry up.

Settlers soon began to ditch and drain floodplain marshes to create rich farmlands. By the 1930s only scattered wetlands remained along Puget Sound rivers. Today few signs other than subtle depressions across plowed fields indicate the former extent of sloughs and marshes that historically provided immense areas of rearing habitat for juvenile salmon. Removing the snags, logjams, and streamside trees that created navigation problems and blocked development eliminated some of the most productive salmon habitat on earth.

The program of clearing wood from rivers did not originate in the Pacific Northwest. Snags and logjams also presented treacherous navigation hazards to steamboats in the midwestern and southern states. In the two decades after the end of the Civil War, Army Engineers pulled 46,858 snags and 651 large drift piles from the Mississippi, Missouri, and Arkansas rivers. During this same period Army Engineers cut down an astounding 180,268 streamside trees along these rivers to prevent future snags. Eventually, close to a million snags were pulled from the lower Mississippi, reshaping the river into a fluvial freeway for moving goods through America's heartland.

In 1886 Captain E. H. Ruffner of the Army Engineering Corps wrote a book to promote clearing logs from rivers in order to improve transportation and commerce. In it Ruffner described the dramatic effects of wood on rivers in regions where such effects are barely acknowledged as possible today. Before engineers cleaned up the South's rivers, snags and logjams obstructed rivers from North Carolina to Arkansas. Although little recognized even in academic circles, the effects of wood in these rivers were similar to those documented a century later in studies of Pacific Northwest rivers.

The most famous logjam, called the Red River raft in Louisiana, was a complex of log dams that filled in 140 miles of river and affected almost 300 miles of channel. The jam probably started forming in the late 1400s and built upstream at an average rate of about a mile per year. By the late 1800s it had advanced from near Alexandria, Louisiana, to within 3 miles of the Arkansas border. The immense jam filled mile after mile of channel and prevented passage of all but the smallest craft.

The water impounded behind the massive jam inundated vast areas of the floodplain, and created huge lakes where tributaries entered the main valley. Captain Ruffner described how bank erosion dropped massive trees into the river, which when anchored into the riverbed would split the flow into numerous side channels. Deposition of sedi-

ment around these massive trees formed elevated surfaces that filled in
to form the valley bottom.

Late in the spring of 1873 Army Engineers finally opened a passage
through the massive river-blocking jam after years of effort and failed
attempts. That same day a boat laden with cotton steamed up the river
and through the opening to inaugurate a new era in river-borne commerce. But success was short-lived. Six years later a flood deposited a
huge amount of wood that once again blocked five miles of the river.
Rising to the challenge, Army Engineers reopened the river the following year. This time they permanently stationed a snag boat on the river.
Learning that rivers remained at risk of obstruction from new logjams
wherever large trees could form new snags led the Army Engineers to
not only remove logs from rivers but to start cutting down trees standing along their banks.

Removal of the Red River raft triggered extensive changes in the
character of the river and valley bottoms. Clearing the log raft drained
lakes and vast wetlands, which by 1900 covered just 3 percent of the
area that they had inundated before the river was cleared. Within a few
decades the farmland created by draining the natural lakes sank beneath the water again, this time under reservoirs created for flood control and recreation.

The consequences of removing logjams in Pacific Northwest rivers
were equally dramatic. In 1982 Jim Sedell evaluated the potential for
loss of salmon habitat from historical draining of valley-bottom wetlands and side channels throughout the Pacific Northwest by studying
pristine rivers in Olympic National Park. His investigation of the distribution of juvenile coho salmon in floodplain channels along the Hoh
and Queets rivers showed that almost all of the young fish were in the
small side channels that crisscrossed the valley floor. Salmon used the
main channel mainly for spawning and migration. Side channels
formed by logjams not only provided much of the coho habitat, they
held almost ten times more fish. Recently, a similar study in the urbanizing Snohomish River basin just north and east of Seattle found that
streams still connected to valley-bottom wetlands host two to three
times more adult coho salmon than streams without wetlands.

Streams flowing through forests support three to four times more coho than rural, urban, or agricultural streams. Thus the clearing of valley bottom forests, urbanization, and agricultural development further reduced the capacity of rivers and streams to support salmon throughout the Puget lowland. Even though valley bottom clearing and agricultural practices on floodplains greatly reduced historical salmon abundance around Puget Sound, regulatory attention to the effects of land use on salmon focuses almost exclusively on upland forestry.

Clearing of small streams and upland tributaries for transporting logs followed a few decades after clearing of major rivers. Getting logs to mills was a major problem for early timber merchants. Railroads were expensive, hard to build, and worth constructing only in relatively flat terrain. Rivers, on the other hand, only needed to be cleared of obstructions in order to be used to move logs downstream. On big rivers, logs floated down to riverside mills or piled up at the limit of tidal influence. On small streams, floods released from temporary dams could sweep logs downstream. This practice, called "splash damming," was imported from Maine, where it originated in the 1700s and continued in use until the last log drive down the Penobscot River in August 1971.

In principle splash damming was easy. One simply blocked the river with a temporary dam, filled the lake behind the dam with logs, and then breached the dam to create a wall of water that swept the logs downstream. The resulting flood would do all the work. In practice, however, splash damming was hard, dangerous work.

Logjams were the enemy on a log drive. Woodsmen armed with peaveys—stout poles with a hook on one end—followed the flood to dismantle any jams that formed. Sometimes these guys actually rode the flood wave, perched in nail-studded boots on top of bobbing logs, knowing that to slip could prove fatal in the churning, log-strewn waters.

For logs to be sluiced downstream by a splash-dam release, the channel needed to be free of large debris, boulders, or bedrock outcrops that could hang up logs shooting down the river. Initially, logjams were dismantled one log at a time, starting with the key logs that anchored the jams. Eventually, companies specializing in log driving decided this really was a job for guys with dynamite. Lots of dynamite.

Scores of rivers were simplified by blasting out inconveniently located logjams, boulders, or outcrops that might impede passage of a log-filled flood.

Thousands of splash dams were constructed throughout the Pacific Northwest. Many rivers hosted a whole series of dams. In coastal Oregon, fourteen dams operated on the Millicoma River between 1884 and the 1920s; fifteen splash dams operated on the Coquille River in the first half of the twentieth century. In coastal Washington, splash dams blocked the entire length of some major tributaries: The Willapa River had seventeen splash dams; the Humptulips River had at least eighteen; and the Chehalis River system had over forty-eight. Repeated splash damming must have been a nonstop catastrophe for migrating, spawning, and juvenile salmon, as repeated filling and breaching of these dams sent flood after flood down scores of rivers and streams. Sustained for decades, such intensive, regionwide disturbance was unlike anything salmon or their streams had ever experienced.

Conflicts between timber and fishing interests started almost as soon as commercial forestry began in the Pacific Northwest. Sawmills needed proximity to markets, a way to get logs there, and flowing water to run saws. The mouths of salmon-bearing rivers and streams naturally provided ideal sites. Sawmills began appearing along Oregon's rivers in the 1830s, and by the 1850s Indians were complaining about mills scaring away salmon. Clatsop Indians living along the lower Columbia River insisted that a sawmill near their village be relocated as a condition for signing their 1851 treaty with the United States. They maintained that the sawmill had ruined their salmon fishing.

The tribes were not the only people to notice the effects of logging on salmon. In coastal Washington, salmon hatcheries were constructed in the early 1900s to mitigate the effects of splash dams on commercial salmon fisheries. In Oregon, concern over destruction of docks, land lost to bank erosion, and effects on salmon fisheries led to the Log Boom Act of 1917, which empowered the Public Utilities Commission to regulate splash damming.

Dow Beckham was a logger who when he retired wrote a memoir, *Swift Flows the River*, about a life spent in the Oregon timber industry.

Beckham relates how a 1940 hearing of the Oregon Public Utility Commission over a splash dam franchise on the Coos River captured the essence of the debate between forestry and fishing interests. The Coos River Boom Company argued that a splash dam was the only way for the company to transport timber from the virgin forest on the South Fork of the Coos River to the sawmills on Coos Bay. Their competitors, the Coos Bay Logging Company, owned the only road along which logs could be transported overland. Coos River Boom argued that the splash dam was needed to promote low-cost log transportation and support the local economy. The Oregon State Fish Commission opposed licensing new splash dams, which they believed would decimate economically important salmon runs and imperil the operation of their salmon hatchery. After a series of acrimonious hearings that spanned several years, the Public Utility Commission ruled that the splash dams would provide the greatest public benefit.

Beckham described how log-driving companies, even when subject to restrictions, simply ignored regulations and were "in almost constant violation of the rules. . . . We could not have handled the volume of logs otherwise. If they had enforced the rules, splash dam logging would have terminated" (1990, 113).

The removal of logs that formerly trapped and held gravel, together with repeated scouring to bedrock during splash dam releases, turned many once-productive salmon streams into bedrock chutes with little value as salmon habitat. A 1950 Oregon Fish Commission report written by J. T. Gharrett and J. I. Hodges on the condition of coastal Oregon streams observed,

> Splash dams. . . built for the purpose of sluicing logs down the rivers, have blocked the salmon runs and eliminated the productivity of the streams above them. This practice has also resulted in the sluicing of the gravel and destruction of the spawning areas below the splash dams. Many impoundments for log ponds, power production, and irrigation have been created without providing the proper facilities for fish passage. These have eliminated the salmon runs in many river sections. (20)

A story in Dow Beckham's memoir conveys how splash damming illustrates two recurring phenomena in the history of salmon. At a public meeting in the 1950s between fishery interests and representatives of the Oregon State Fish and Game commissions and the Menasha Corporation, which maintained a splash dam, everyone in attendance except the company officials agreed that Menasha's two splash dams provided inadequate fish passage. Even so, representatives of the state wildlife agencies maintained that no rules or regulations were being violated. Positions staked out at this meeting illustrate how industries contributing to the decline of salmon steadfastly deny even obvious impacts. At the same time, regulatory agencies charged with translating the intent of laws into rule and regulations in fact accommodate powerful economic interests when they shape the rules and their enforcement.

The splash dam era in Oregon spanned eight decades, starting in the 1880s, when timber harvesting began to spread from river valleys into upland areas. Within a decade, a local newspaper, *The West Shore*, reported that in Columbia County, Oregon, every "stream of any size has been cleared of obstructions, so that logs can be run down them in the high water season" (Sedell et al. 1991, 326).

By the turn of the century, the same could be said for counties throughout Puget Sound, the lower Columbia River, and coastal Oregon and Washington. Over three hundred major splash dams operated year-round. Many more operated seasonally on small streams and tributaries. In some upland streams, splash dams operated for just a few decades before the practice began to wane as logging roads and trucks replaced streams and railroads as the logger's transportation network. In the mid-1950s, once river transport of logs was no longer essential to the powerful timber industry, the Oregon state legislature banned the practice, in 1956. In 1966 Potlatch Timber conducted the last log drive in the Pacific Northwest, down Idaho's Clearwater River.

That any salmon at all remain in streams with a legacy of splash damming testifies to their resilience in the face of disturbance. In a pamphlet written for the Washington State Department of Fisheries, *Logging Dams on Coastal Washington Streams*, H. O. Wendler and

G. Deschamps described the effects on salmon of the splash dams operating on coastal streams:

> If fish were spawning, the sluiced logs and tremendously increased flows would drive them off their nests. . . . After splashing, no fish were seen, nor were any seen the following day.
>
> Besides harming the fish, physically, the stream environment was often adversely affected by splashing. Moving logs gouged furrows in the gravel and in many instances the suddenly increased flows scoured or moved the gravel bars, leaving only barren bedrock or heavy boulders. . . . If the sudden influx of logs into the stream below the dam caused a log jam, as was frequently the case, dynamite or black powder was used to clear the obstruction. The policy followed in those days was that if examination showed that two boxes of powder would suffice, four were used. . . . Great numbers of salmon and steelhead trout were reportedly killed by these blasts. (Wendler and Deschamps 1955, 31–32)

The effects of industrial logging on salmon runs were also noted in British Columbia. In *Fishery Problems in British Columbia,* John Laws Hart of the Pacific Biological Station in Nanaimo, Vancouver Island, concluded,

> [H]arvesting of the timber resources in British Columbia has threatened the prosperity of the fisheries for the great anadromous species. . . . [T]he effects of stream deterioration are so great that it is questionable whether it is possible now to state what part of the decline in abundance of salmon is due to interference with spawning grounds and what part of the decline is the result of the fishery itself. (1950, 422)

By midcentury it was clear that habitat degradation due to logging operations was a major factor contributing to depletion of salmon runs on the Pacific Coast.

The pace of landscape change accelerated in the second half of the twentieth century. In short order the newly invented chainsaw and bulldozer reshaped the practices of the U.S. Forest Service and the timber industry, and the land itself. Trees could be felled much faster than before, and roads could be punched in almost anywhere. After the Second World War, the U.S. Forest Service and the timber industry constructed in the nations forestlands the most extensive road network ever built. Almost half a million miles of logging roads form a system almost ten times more extensive than the interstate highway system— long enough to circle the globe more than fifteen times. The length of the road system rivals or exceeds the total length of streams and rivers in many areas across the forestlands of the Pacific Northwest.

Once, when I asked a timber industry colleague about access to a potential study site, he jokingly assured me that we didn't need to worry. His company had roads to within 100 yards of anywhere on their vast holdings. His jest isn't far from the truth. The pace of timber cutting increased astronomically as logging roads carved the landscape into forest patches about the size of city blocks.

Almost immediately, fisheries biologists noticed an effect on salmon runs. Gharrett and Hodges's 1950 report, in which they summarized the effects of logging on coastal Oregon streams, concluded that such effects were obvious. Specifically, they found that logging operations resulted in loss of gravel from streambeds, silting in of spawning gravels, and higher stream temperatures. Their report also noted how branches and unmarketable wood dumped into streams by logging operations resulted in logjams that create deep pools thought to be impassable to fish, and they were careful to discriminate between most natural logjams and those that were impenetrable fish barriers.

Concern over the effects of logging slash in streams and the blocking of fish passages led biologists to call for stream-cleaning programs to keep logging debris out of streams. Fisheries biologists made it a priority to remove such jams to open up access to hundreds of miles of spawning areas. Growing perception of logging-generated logjams as barriers to salmon migration set the stage for new forest-practice rules prohibiting the dumping of logging debris in streams.

Soon, however, the idea of keeping logging debris out of streams mutated into a paradigm of keeping rivers and streams free of wood altogether. From the 1950s to 1970s natural resource agencies worked diligently to clear wood from streams. The landmark 1972 Oregon Forest Practices Act actually required removal of wood from streams. Washington adopted similar rules. California too. All along the West Coast a clean stream not only looked like a good idea, it was the law. Ironically, recognition of the cumulative effects of many small man-made dams in the decline of Atlantic salmon provided a (misguided) rationale to strip natural logjams from Pacific Northwest streams. Over the next several decades, increased understanding of Pacific Northwest streams and rivers revealed that this notion was tragically misguided.

Not everyone advocated stripping all wood from rivers. Henry Froehlich, a professor of forest engineering at Oregon State University, cautioned, "Overcleaning channels may not be in the best interest of stream protection. We must recognize what the stream systems are like in their natural condition and also to understand the role of the organic debris in these channels" (Froehlich 1973, 84).

At first Froehlich's advice fell on deaf ears. Without the ability, training, or inclination to determine which wood was natural and which was logging debris, most stream-cleaning crews simply removed everything they could lift or hoist from channels. Soon it became difficult to find streams that had not been stripped of logs and logjams.

Following Froehlich's call to base stream management on an understanding of natural stream processes, scientists learned that large woody debris is a major influence on stream ecosystems and salmon habitat in the Pacific Northwest. Current understanding of Northwest salmon streams holds that formation of logjams, channel movement during floods, and catastrophic disturbances create and maintain some of the best salmon habitat. Sloughs and side channels that harbor juvenile salmon are created when rivers jump around their valley bottoms as a result of floods or logjams. Big logs that fall into rivers and streams help form deep pools and provide structure to salmon habitat. Curiously, destructive landslides and floods that knock big trees into streams and rivers help provide an environment in which salmon thrive in the times between such disturbances.

Changes in land use and forest cover can alter how sediment and water move across and through the landscape and thereby change the structure and dynamics of salmon habitat. The Skokomish River, which flows into Puget Sound east of Olympia, is a poster child for the impacts of misguided forest practices, river management, and land use in the Pacific Northwest. Normal rivers spill over their banks about once a year or two. That's normal. In most years along most rivers, the annual flood is not much higher than the banks and causes little inconvenience even to those living on the floodplain. Yet the Skokomish River now floods many times in most winters, leaving no doubt that it is one seriously disturbed river.

Year after year Seattle's TV newscasters diligently report that the Skokomish is once again flooding at the first sight of rain without questioning or investigating why this is happening. In the winter of 1998–99 the Skokomish River flooded twelve times. Mason County emergency management officials complain that it seems the county has to respond to the threat of flooding almost every time it rains. Rivers simply are not supposed to flood several times a month all winter long. Bill Hunter, a seventy-two-year old farmer who had lived along the river most of his life, complained that the river now floods so frequently that it limits what he can grow. He recalls when the river had deep pools, in striking contrast to the present shallow, featureless channel.

So what caused the mess on the Skokomish? In 1929 the city of Tacoma's Cushman Dam was built on the North Fork of the Skokomish River, which reduced the flow in the river downstream of the confluence of the North and South Forks. This also reduced its ability to carry sediment. Intensive timber cutting destabilized steep slopes in the basin headwaters, and gravity carried hill-slope soils down to the river. The level of the riverbed started to rise as the slug of sand and gravel moving down the South Fork arrived at the confluence with the de-watered North Fork. The supply of sediment moving downstream simply overloaded the river, which filled in 6 feet between 1950 and 1990. The rain hasn't stopped, the runoff still has to go somewhere, and the channel can't hold it anymore. So now the river frequently spills out over its banks and across the floodplain.

The Skokomish River is a classic example of "living downstream." Yet in all the TV time dedicated each year to covering flooding on the

In November 2001 a chum salmon tries to get across a flooded road to reenter the Skokomish River.

Skokomish, nobody talks about how upstream land use affects folks living downstream. The people living on the floodplain of the Skokomish River did not build the dam. They don't receive power from the dam. They did not profit from cutting the trees down in the basin's headwaters. They just got stuck with the bill: houses, farms, and roads that get flooded all the time.

The effects on salmon populations of changing stream environments became increasingly obvious to fisheries managers in the second half of the twentieth century, even though the basics were understood fifty years ago. In 1951, even as their annual reports continued to blame Indian fishing for declining salmon runs, the Washington State Department of Fisheries published *The Salmon Crisis,* a pamphlet which stated frankly, "The main causes of salmon depletion can be traced directly to the environmental changes that have taken place since the advent of civilization in the Pacific Northwest" (1951, 5).

Although changes in habitat structure due to loss of large trees transformed the physical structure of Pacific Northwest rivers, other environmental and land-use changes played an important role. Agriculture spread rapidly across floodplains, further simplifying channels. Within a few decades after rivers were cleared of snags and logjams, the flood-plain sloughs were ditched, valley-bottom wetlands were drained, and side channels were plugged. Increased human occupation and use of floodplains made the natural occurrence of flooding a problem, so extensive networks of dams and levees were built to control floodwaters. Many rivers were straightened to reduce flooding and speed water on down the river. After a twentieth-century orgy of dam building, the continental United States now has 75,000 dams that together can store an average year's runoff. An estimated 25,000 miles of levees enclose more than 30,000 square miles of floodplain in the United States, an area larger the combined area of Massachusetts, Vermont, and New Hampshire.

Extensive construction of levees and dikes diminishes floodplain storage of water during floods, for confining the water within a walled-in channel just pushes flooding farther down the river. So once levees are built along part of a river, pressure mounts to extend the system downstream until eventually levees line the whole river. The result is a straight-jacketed river that can no longer move across its floodplain. Once locked in place, a river no longer supports natural processes of channel migration that create the side channels and off-channel water bodies that shelter young salmon.

Flood-control efforts can also exacerbate flood hazards by encouraging human occupation of flood-prone areas. Flooding is a problem created by people living in or building on areas that become part of the river during high flows. Settlers knew that rivers periodically spilled over their banks, so they built on high ground in the valley bottoms, above the level of frequent inundation. Once levees along a river stop the annual high flows from reaching the floodplain, development typically spreads across the flat ground near the river—usually right up to the levee. Unfortunately, rare but inevitable heavier flooding puts floodplain development at risk when the water rises over the levees or they eventually fail. I doubt whether old-timers, if they had the chance, would buy houses in new developments sprawling indiscriminately across floodplains, especially those in drained wetlands lying at (or sometimes even below) river level.

New development, levees, and fields on the floodplain
of a river near Seattle in the 1990s.

As a piece of the landscape, floodplains occupy a relatively small percentage of the mountainous Pacific Northwest. In the region around
Puget Sound, direct and indirect subsidies maintain human occupation
and uses that run headlong into the inevitable periodic inundation of
floodplains. Such subsidies include public money spent on rebuilding
dikes and levees in the same places over and over again after floods.
Additional costs accrue for repairing bridges, houses, roads, and businesses on flood-prone land. Policy makers and agencies are starting to
ask why society subsidizes, and thereby encourages, human occupation
of floodplains. Still it remains rare for the loss of salmon to be included
in assessments of the costs associated with floodplain development.

In a 1964 report, "King County, Washington, Comprehensive Plan for Flood Control," written for the Board of King County Commissioners on flood protection along Seattle's Cedar River, the engineer Bertram Thomas noted, "The stream historically has claimed the whole valley for its use. . . . As a matter of fact, the flood control funds disbursed on the Cedar River in recent years would have covered acquisition costs of all lands along the river several times over" (1964, 26–27). In other words, the cheapest long-term solution to flooding problems would be to simply buy the entire floodplain and relocate the people living there. Out of step with the times, this unconventional idea was ignored. Since then, development that spread across the floodplain greatly increased the cost of flood control, and the river was locked into place by levees. In recent years the county has spent more on restoration projects along the Cedar River than it would have cost to simply buy the entire floodplain in the early 1960s.

Around Puget Sound, urbanization changed the way water moves across and through the landscape. Pavement and buildings prevent water from sinking into the soil. This increases runoff in proportion to the area of impervious surfaces. In many developed areas, the increased runoff turned the annual high flow into an annual cataclysm for stream-dwelling organisms. Not surprisingly, fish with eggs in the gravel during the months when storms normally hit quickly disappeared. Adapted to digging nests in tune with smaller flows, the higher flows scoured out their incubating eggs. The typical pattern for urban streams was for fall-spawning salmon to be replaced by cutthroat trout, which spawn in the spring when their eggs are generally safe from high flows.

The Washington State Department of Fisheries recognized the threats posed by urbanization, stating in its 1965 annual report, "Real estate development continues to contribute an increasing number of hydraulic project proposals . . . but protection to the fishery is not always easily attained as fishway construction and alteration of plans may be costly to the developer" (Washington State Department of Fisheries 1965, 61). Project by project, development modified streams across the Puget Lowland. Powerless to stem land-use changes, the state Department of Fisheries lacked the authority and mechanisms to enforce meaningful salmon protection measures in the rapidly urbanizing Puget Lowland.

In the ten-year plan the Department of Fisheries prepared for the state legislature in 1967, it offered a stark prognosis.

> If no more stream losses were to occur, natural production of salmon would probably be maintained at a relatively high level. This, however, will not be the case, as evidenced by recent initiation of extensive channelization and flood-control work. . . . The future outlook for Puget Sound and Washington rivers and streams for salmon use appears rather bleak if we allow the present trend of hydraulic changes and encroachment to persist. (Washington State Department of Fisheries 1967, 25–26).

Although no directives from agency heads and elected officials privileged developer's interests over salmon, the system was set up so that the fish would lose—one project at a time.

In addition to paving over or changing the hydrology of streams, urbanization also drove demand for gravel, which is essential for construction and typically is mined from riverbeds. Surprisingly, the annual value of sand and gravel mining exceeds by far the value of gold mining in Washington State. As Washingtonians poured more and more concrete, gravel mining carved deeper into riverbeds and farther across floodplains along major rivers. Unfortunately, gravel is also essential to salmon, as the Department of Fisheries noted: "Gravel removal is a constant and increasing problem. Gravel is a basic element to natural salmon reproduction and excessive removal . . . can be extremely costly to the salmon resource. The impact in many cases can never completely be overcome. . . . Little change in this trend is envisioned in the near future" (Washington State Department of Fisheries 1965, 60).

Development also fueled demand for water. Projections of the water demand for industry and agriculture in the 1960s left "little doubt that many streams presently having adequate minimum flows for fishery purposes will be in serious trouble within the next decade" (Washington State Department of Fisheries 1966, 176). Thus, as urbanization proceeded across the Puget Lowland, it put people and salmon in direct competition for the two key things salmon need most: water and gravel.

Urban streams present a difficult challenge for salmon recovery efforts because of the fundamental changes in runoff and land cover that accompany development. Some measures to manage urban runoff, such as using porous concrete to allow rainfall to sink into the ground and thereby slow runoff, may be expensive at first but economical if considered over the long haul. Many other measures simply require not doing certain things that developers and homeowners like to do, such as paving over driveways with hardtop, clearing streamside trees and armoring streambanks, or building levees to keep channels from moving around. Simple measures for rehabilitating urban streams include managing stormwater to prevent excessive increases in peak runoff, protecting natural wetlands from development, moving levees back from the channel's banks, and prohibiting development in "channel migration" zones so that channel movement is possible across a portion of the river's floodplain.

In addition, we must decide which rivers and streams to protect, and which if any to write off in the urban environment. This point is controversial and is rejected by many committed environmentalists. But the Port of Seattle is not going away anytime soon. The habitat quality of the lower Duwamish River is improving, but it is simply not restorable in any reasonable time frame to its former glory as a salmon river. In contrast, the relatively rural Snoqualmie River and many of its tributaries still can be protected from urban encroachment and development. Given the projected doubling of the human population around Puget Sound in the next fifty years, choices need to be made, or they will be made by default and accident rather than by design. And it was the day-to-day accumulation of little decisions that over time transformed the freshwater world of salmon.

Recognition of the effect of forest clearing on Pacific Northwest rivers is leading to regional efforts to reintroduce wood to rivers. Controversy over allowing logs and logjams back into rivers currently focuses on the interests of recreational boaters who want to keep rivers clear of logjams, which can present a hazard to kayaks and canoes. Though boating organizations certainly are justified in seeking to minimize risks for boaters, a perfectly safe river free of wood is a river with few salmon.

The size of forest buffers to be left uncut along streams also is controversial in both commercial forestlands and urban and rural lands. Until recently the debate over the size of forest buffers neglected the key point that rivers move. But now channel migration zones are being prescribed to account for the propensity for channels to move around across a river's floodplain. Most regulators understand that wood is important to fish, but I suspect that few fully comprehend what will happen as big trees are reintroduced as a dynamic element in rivers. As big wood returns, rivers will start to once again behave as they used to, bouncing around their floodplains and eroding their banks and depositing bars of sediment. In a pristine river this is the norm. But now there are people living on floodplains. And there are roads and towns and other things in the way that people don't necessarily want to lose to a migrating channel. If we succeed in restoring rivers they will definitely behave differently from what we are used to, and will potentially cause the kind of trouble that salmon could use for a change.

A key question for salmon-recovery efforts in the Puget Sound region is where we want our rivers to fall on a continuum from a wood-rich, multichannel river to a controlled, wood-free canal. Should we target habitat conditions as they were in the 1850s at the time of the Indian treaties? Or should we strive for conditions in 1900 immediately after the clearing of valley bottom forests? It would be easier, and less effective, to set the bar a little lower and target conditions in the 1950s after rivers were canalized, levees blocked off of side channels, and valley bottom wetlands were drained. The easiest, and least effective, approach would be to hold the current line and guard against *further* degradation of salmon habitat.

Naturally, the optimal strategy depends on whom you ask. If you give top priority to navigation or development in the floodplain, the wood-free, levee-bound Snohomish River is the obvious model. If you give top priority to salmon production, then you are aiming for a river like the wood-filled, split-channel lower Nisqually River. But salmon-recovery goals should also take into account the innate capacity of the river to produce salmon. Unfortunately, land-use changes have caused rivers with historically very different channel patterns to lose their distinctive features and assume relatively similar, simplified patterns.

The impacts of freshwater habitat loss are not evenly distributed among salmon species. Chum salmon (*Oncorhynchus keta*) populations around Puget Sound are doing well in spite of the dramatic changes to river systems. More or less ignored by commercial fishermen, they've been increasing in recent years. Pink and sockeye runs are also doing reasonably well, more or less holding their own. It is the chinook and coho runs that are in the greatest danger, with populations continuing to decline below perilous levels.

Why the big difference? Chum salmon spend very little time in freshwater. Juveniles pretty much head straight to sea after hatching, minimizing their dependence on freshwater habitats. In contrast, coho and chinook spend one or more years in freshwater habitat before heading to sea. It is no coincidence that the salmon most dependent on freshwater habitat are in the deepest trouble. Conversely, it is no miracle that the Queets and Nisqually rivers, which remain much as they were before civilization transformed the region, still support relatively healthy runs.

Recovery of Puget Sound coho and chinook will only happen if we let rivers behave more as they did when they sustained large salmon runs. These fish not only need water, they need dynamic rivers to create and sustain suitable habitat. Without a doubt, the best way to allow a river to retain some natural character is to give it a little room to move around and, well, act like a river.

THE SIXTH H

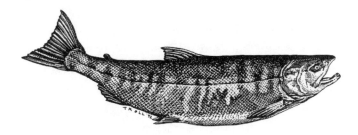

Our present problem is one of attitudes and implements. We are
remodeling with a steam shovel, and we are proud of our yardage.

Aldo Leopold, *A Sand County Almanac*, 1949

THE THOUSAND-YEAR RUN OF SALMON ILLUSTRATES HOW
technological advances and land-use change can magnify the effects of intensive fishing to eventually exhaust a public good. Details
vary from region to region and from river to river. But recurring themes
run through the strikingly similar stories of salmon declines throughout
the animal's range.

Huge salmon runs initially provided an abundant food source that
sustained subsistence economies in Europe, the American-Canadian
Northeast, and the Pacific Northwest. In all three areas, native people
relied on salmon fisheries, which they protected through cultural practices that restrained overexploitation. In the absence of technologies to

preserve salmon or markets for export, fishing intensity matched the modest needs of local consumption. As long as local human populations depended on local salmon there was a built-in ecological safeguard. People who overfished, or otherwise degraded their fishery, cut off their own life support.

In the thirteenth century, Edward I of England had no scientists, no professional fisheries managers, and no industry lobbyists to contend with. Yet he could see clearly what was needed to protect his country's salmon—access to their habitat, an open way to the sea, and fishing practices that did not overexploit individual runs. Mandates to enforce this vision worked for hundreds of years until commercial fishing and industrial interests began to influence legislation as industrialization and urbanization transformed the landscape.

In the New World no authority was willing to limit entrepreneurial access to salmon and the populace increasingly ignored laws and traditional cultural restraints as immigrants less dependent upon salmon displaced native people. At first, an abundant supply of salmon helped feed the lower classes, and farmers considered salmon cheap fertilizer or livestock feed. Once merchants were able to preserve and ship salmon to distant cities, they quickly sent rivers of fish streaming to market. The accelerating pace of technological development compressed the time frame over which these market and cultural developments occurred, from centuries for English salmon to one century for the salmon of New England and Canada to decades for the Columbia River runs. But in each region the original, publicly held resource became an overexploited commodity when technology for preservation and shipping made exporting salmon commercially viable.

The transformation of the salmon's habitat into farms, towns, and cities amplified the detrimental effects of overfishing by reducing the capacity of rivers to support salmon. Clearing of logs and logjams, as well as channel straightening, diking, and damming for flood control degraded salmon habitat in river after river. Already stressed salmon populations crashed as forests were cut and dams blocked rivers and streams. In many areas, pollution from industrial and urban wastes polished off most, if not all, remaining salmon. Finally, urbanization converted some channels into little more than inhospitable concrete-lined ditches. As these changes progressed, salmon-conservation efforts failed over and over again as steps to deal with one set of factors were undermined by other causes and a general lack of will to enforce laws and regulations.

The English, in transforming the economic profile of their island, sacrificed their salmon for the modern age of the industrial revolution. New Englanders, too, traded salmon for a tamed landscape better suited to support their agricultural needs and industrial aspirations. The feverish rush to extract gold from riverbeds destroyed much of California's salmon. Even before the promise of cheap water and electricity drove the construction of dams that impeded the migration of Columbia River's salmon between the sea and their spawning grounds, most of the Columbia River's huge chinook had already been canned and shipped east. At the same time, the Pacific Northwest's ancient forests, which structured salmon habitat, were cut over and converted to timber plantations. Unscreened irrigation diversions sucked not only water from rivers but also young salmon into farmers' fields. In each chapter of this saga salmon habitat or salmon themselves generated capital that helped finance regional development and placed further pressure on wild salmon.

Faith in the ability of hatcheries to produce more fish allowed overfishing and degradation of rivers to continue without acknowledgment that the loss of salmon was the price of unrestrained development. Modest initial successes of hatchery programs fueled a second round of commercial overharvest. For a time, hatchery fish sustained large commercial fisheries. But instead of rebuilding spawning runs, hatcheries propped up shrinking populations by pumping out smolts ill equipped to survive in the wild. In the end, reliance on hatcheries replaced wild salmon with hatchery fish, and delayed but did not reverse the ongoing decline in salmon abundance.

Once technology enabled intensive open-ocean fishing, even countries that never had, or had already lost, salmon from their rivers, such as Denmark and Japan, could chase salmon out at sea. With this new pressure, salmon populations continued to plummet and whole fisheries collapsed. As the price of salmon rose, fish farming became not only profitable, but so successful that farmed salmon now outnumber the wild salmon in Europe and New England. They are gaining on wild fish in the Pacific Northwest.

The end result of these recurring chapters has created a story in which once-common salmon are becoming or have already become rare. Salmon returns to Pacific Northwest rivers are just 6 to 7 percent of historic levels. Failure to learn the lessons of past experience is leading to a familiar outcome—the exhaustion of another region's salmon runs. Despite rhetoric to the contrary, modern management of salmon and their habitat provides a superb example of maladaptive management—the failure to learn by experience.

Salmon are resilient, robust animals that can rapidly colonize new environments. They are more like weeds than like a sensitive bird that can only nest in a special type of tree that occurs in a particular type of forest in a couple places on earth. Even so, we are managing to drive them to the verge of extinction across much of their range.

The Pacific salmon have disappeared from a third of the area they inhabited just 150 years ago in California and the Pacific Northwest. In 1991 three widely respected fisheries scientists reported on the status of distinct populations, or stocks, of Pacific salmon in the continental United States in "Pacific Salmon at the Crossroads: Stocks at Risk from California, Oregon, Idaho, and Washington" (Nehlsen, Williams, and Lichatowich 1991). One third of Pacific salmon stocks were already extinct and half of all surviving stocks faced a high risk of extinction, they wrote. Only one out of six of the original salmon stocks in the Pacific Northwest remained in good shape—neither extinct nor at significant risk of extinction. In the decade since this bleak assessment, salmon runs have been listed under the Endangered Species Act (ESA) in all four states.

Years after the first ESA listings of salmon, goals for salmon-recovery efforts remain vague. State and local governments generally prefer not to define their own goals for fear that they may set more stringent objectives than the National Marine Fisheries Service may ultimately establish. The lack of well-defined strategic objectives undercuts meaningful salmon-recovery efforts and allows important factors acknowledged to influence

Cover illustration for R. D. Hume's *Salmon of the Pacific Coast*.

their decline (such as floodplain development and water diversions for agricultural purposes) to remain virtually off limits in the development of policy options. Instead, salmon-recovery goals remain operationally defined by whatever is easiest to do with little substantive change to business practices, lifestyles, or regulatory and enforcement policies.

More troubling is that elements of current salmon-recovery efforts remain at odds with basic lessons from past salmon crises. One of the most obvious lessons of past experience is that local control rarely protects salmon over the long run without direction from a higher authority, whether the king, a federal agency, or, as for Native Americans, the Creator through deeply ingrained cultural practices. Yet today, faith in local control provides the basis for salmon-recovery plans, and the federal agencies charged with enforcing the ESA are rarely aggressive in either guidance or enforcement efforts. Another lesson is that grandfathering of existing impacts in formulating regulations will, over time, progressively deplete the capacity of rivers to support salmon. Recent efforts by some state legislators to protect existing water uses based on questionable water rights that are threatened by ESA listings of endangered salmon show that this lesson remains unlearned.

A parallel problem is that even gradual habitat loss and degradation eventually add up to substantial impacts. Other than the obvious case of impassable dams, individual projects rarely drive salmon from a river. Instead, lots of small changes add up to do so over time. Highly visible symbols like dams are obvious targets for salmon advocates, but the net effects of less obvious changes that incrementally alter the landscape are more insidious and difficult to remedy. It is particularly hard to control the gradual reshaping of the landscape by the kinds of land use that we, as a society, permit and even encourage. Although decades are a long time for societal planning, longer-term trends can determine the success (or failure) of salmon-recovery plans. A long-term plan that lays out the vision for defining recovery and a strategy—a roadmap—for achieving it is needed to ensure sustained progress.

In principle, crafting such a plan should not be difficult. Salmon need just a few basic things:

- Cool, unpolluted water.
- A clean gravel streambed that doesn't scour out or entomb their eggs in deposits of fine sediment.
- A flood regime in tune with their life cycle.
- Accessible habitat that provides food and cover from predators, as well as areas where juveniles can grow and develop before heading to the ocean.
- A chance for enough juveniles and adults to evade open-ocean and river fisheries so that they can return to their home rivers and spawning beds.

Given these things, salmon can thrive. Without them, they will eventually disappear.

For over two centuries now, reduced fishing pressure and removal of barriers to salmon migration have been proposed over and over again as essential to preserving or restoring salmon fisheries. In addition to these commonsense measures, an effective strategy for long-term salmon recovery would need to include the following actions:

- Protect high-quality habitat and aggressively enforce existing regulations to prevent degradation of critical salmon habitat.
- Restore rivers and streams, guided by an understanding of historical conditions and the salmon-producing capacity and potential of individual rivers.
- Reform hatcheries so their focus changes from serving as "salmon factories" to rebuilding wild runs.

According to these criteria, most regional salmon-recovery initiatives inspire little confidence that they will lead to the recovery of self-sustaining runs of wild salmon. Yet there are examples where these simple principles have been applied successfully.

By the early 1990s the winter run of Sacramento River chinook had shrunk to under two hundred fish, and the National Marine Fisheries Service listed the run as threatened under the Endangered Species Act. This move catalyzed a broad cooperative effort by environmental groups and state and federal agencies to address some of the long-acknowledged causes of the decline. Two large irrigation districts and state water projects in the Sacramento delta installed modern fish screens on water intakes. Pumping schedules at state water project intakes in the Sacramento River delta were changed to reduce impacts on migrating smolts. Dam operators at the Red Bluff diversion dam also altered their practices to accommodate the downstream passage of migrating smolts. Obsolete dams on Clear Creek and Butte Creek were even removed, reopening habitat lost decades before. At the Livingston Stone Hatchery (named for the early advocate of artificial breeding), located at the foot of Shasta Dam, practices were reoriented toward augmenting natural reproduction. A decade later, three thousand to seven thousand winter-run chinook were returning to the river, and the population was increasing. When commonsense solutions are aggressively pursued and enforced, the downward spiral in salmon numbers can be reversed.

Where salmon recovery—rather than just a slower rate of decline—is the goal, there must be no further increase in the net impact of the four H's— habitat, harvest, hydropower, hatcheries. How can a strategy reflect this requirement and still accommodate the needs and economic desires of a growing human population? What can be done to reverse current trends and increase salmon abundance around Puget Sound in the face of the projected doubling of the region's human population in this new century?

Currently, hatchery reform remains controversial politically. But some progress is being made with the first H, habitat protection and river restoration. There are a number of additional measures that could be put in place to help recover salmon. Three ideas that would help current salmon recovery plans include:

- Establish independent riverkeepers with the authority to either enforce or trigger enforcement of laws, implement local recovery efforts, and coordinate local, state, and federal actions.
- Establish salmon sanctuaries on valley bottoms and floodplains to create salmon-friendly environments that also can be enjoyed by people.
- Establish a five-to-ten-year moratorium on fishing for at-risk species and then restrict fishing intensity to no more than half of any run.

The lessons of history and well known factors that lie behind this three-pronged approach include the difficulty of local enforcement, the sensitivity of rivers to the cumulative effects of gradual landscape change, and the basics of salmon biology.

Almost any action to help salmon will inconvenience some people and benefit others. Particularly on controversial issues, any consensus that satisfies all stakeholders will ultimately sell out the salmon. So reliance on local control and voluntary measures needs to be guided by an overriding strategy that is guaranteed and enforced by a higher authority. To ensure the success of an overall recovery plan, that authority has to be able and willing to evaluate progress and enforce salmon-conservation measures.

Saving salmon will require changes in how we use land and rivers (not to mention the oceans). This cannot be done on a spot basis. Although it is a fundamental responsibility of government to exercise or impose controls against natural-resource depletion, the public and government must share this responsibility and work together. Somehow we have to align the actions of the many agencies, authorities, and interests that influence salmon streams. At the least, some entity or individual vested with broad authority needs to ensure that the actions of one agency or interest do not undermine broader salmon recovery goals.

An updated version of England's original model of riverkeepers could provide a model for coordinating the vast array of overlapping jurisdictions, tracking resource conditions, and looking out for the interests of society at large (and those of the salmon). As a voice for the resource, stewards charged with protecting rivers and the salmon in them could help prevent or mitigate incremental degradation of salmon habitat. They could ensure enforcement of laws and regulations, such as those for fish-passage and water diversion, that have been on the books for a long time but remain mostly unenforced. Coordinating their efforts through a regional authority and vesting these individuals with the power to write up transgressors and haul them into court would go a long way toward protecting salmon and salmon habitat. Most important, river stewards can provide a sustained focus for institutional memory and help to inform and educate the public about the reasons for, and rationale behind, rules and regulations intended to foster salmon recovery.

No one actually sought to destroy the salmon runs of Britain, New England, and the Pacific Northwest. Decimated salmon runs resulted from individual and collective greed, negligence, and indifference, together with the cumulative impact of human activities that gradually degraded the ability of the land to sustain salmon. What relevance does this legacy have for today's political landscape ruled by the mantra of stakeholder involvement?

The success or failure of current recovery efforts will depend on whether salmon are recognized as equal stakeholders in their own survival, or as junior partners whose interests are measured against those of powerful and influential economic interests. As we try to devise an effective long-term strategy for salmon recovery, it makes sense to examine the effectiveness of past experiments with local control for salmon management. The record is not good. In New England, acts passed by the colonial legislature as early as 1741 provided for inspection of dams to ensure the adequacy of fish passage. Enforcement was provided for at the local level but did not take place, and owners of mills and milldams locked salmon out of one river after another. In the Pacific Northwest, states consistently privileged local economic interests such as the timber industry over the long-term productivity of salmon runs. Only in Alaska, where until recently the federal government *was* the "local" jurisdiction, were regulations to maintain the fishery effectively enforced. Clearly, salmon need advocates to stand up for them against competing interests.

But advocates aren't enough. Salmon need refuges where they are protected from the need to compete with other interests, strongholds where some runs can thrive and not just hang on until random disturbances knock them out, or conflicts arise wherein a river steward may be able to step in and defend the fish. The compelling need to provide salmon with sanctuaries where their interests would not be in direct competition with human interests has been recognized for a long time. Over a century ago, Livingston Stone, renowned agent of the U.S. Fish Commission, reflected on the history of Atlantic salmon and forecast the decline of the Pacific salmon. In "A National Salmon Park," a speech delivered to the American Fisheries Society in 1892, Stone pleaded for the establishment of salmon preserves:

> I will say from my own personal knowledge that not only is every contrivance employed that human ingenuity can devise to destroy the salmon of our West coast rivers, but more surely destructive, more fatal than all is the slow but inexorable march of those destroying agencies of human progress, before which the salmon must surely disappear as did the buffalo of the plains. . . .
>
> To substantiate this statement, which may seem exaggerated, let me inquire what it was that destroyed the salmon of the Hudson, the Connecticut, the Merrimac and the various smaller rivers of New England,

where they used to be exceedingly abundant? It was not overfishing that did it. . . . It was the mills, the dams, the steamboats, the manufacturers injurious to the water, and similar causes, which, first making the streams more and more uninhabitable for the salmon, finally exterminated them altogether. . . .

Provide some refuge for the salmon, and provide it quickly, before complications arise which may make it impracticable, or at least very difficult. . . . If we procrastinate and put off our rescuing mission too long, it may be too late to do any good. After the rivers are ruined and the salmon gone they cannot be reclaimed. Exaggerated as the statement seems, it is nevertheless true that all the power of the United States cannot restore salmon to the rivers after the work of destruction has been completed. (Stone 1892, 150–51, 160–61)

Responding to Livington Stone's call for a salmon park, President Benjamin Harrison on Christmas Eve 1892 created a salmon refuge on Alaska's Afognak Island and Uganik River. It was the only such preserve ever established in the United States outside of the national park system. However, the original fishing restrictions were abolished with the passage of the White Act in 1924 and the reserve lost federal protection in 1959.

Stone was not alone in seeing the need to provide salmon a few safe places in which to weather the changes sweeping across the land. In his 1895 report on how to restore the Pacific Salmon fishery, U.S. Navy Commander J. J. Brice echoed Stone's call for a salmon sanctuary. In 1912, the biology professor Dr. Henry Ward also called for creation of salmon preserves. Twenty years later, the Washington State supervisor of fisheries, Charles Pollock, wrote that the need for "definite legislation mandatory on all state departments involved to set aside certain watersheds as permanent fish sanctuaries to guarantee both commercial and recreational fisheries of this state in the future" was an issue of the "gravest concern" (1932, 34). In 1959, Ross Leffler, the Assistant Secretary of the Interior under President Eisenhower, proposed making the Snake River a salmon refuge to safeguard against the effects of dams and agriculture in the Columbia River basin, but once again the effort failed.

Protection of whole watersheds as salmon preserves still makes a lot of sense, and is physically (if not politically) possible in a place like Alaska, where habitat remains relatively pristine. In the Puget Sound area, undeveloped watersheds are hard to find and people are an integral, growing part of the landscape. Given the inevitable growth in the human population living around Puget Sound, we need a more flexible idea than simply dedicating whole drainage basins as salmon preserves.

How could we accommodate more people and still allow salmon populations to rebound? Through focusing on creating a series of greenbelts along major river valleys we can protect the places that are vital for endangered salmon while also protecting people and property from hazardous, damaging floods. This is a sensible thing to do, because wide valley bottoms typically coincide with floodplains where houses and businesses are frequently battered during floods. Many land uses harmful to salmon are maintained on floodplains only through massive societal subsidies for flood control, levee maintenance, emergency response, and post-flood reconstruction. Can we really afford to continue, let alone support, practices harmful to both salmon and ourselves?

Defining adequate channel migration zones and buffers in urban, rural, and agricultural areas could dramatically reduce federal emergency response dollars spent rebuilding infrastructure such as roads and levees over and over again in the same places after floods. Restoring the ability of rivers to move around and reestablish side channels together with reopening connections between rivers and off-channel wetlands would go a long way toward restoring salmon habitat. Although letting rivers and floodplains revert to a more dynamic state may sound radical, it is worth noting that historically these very same areas were nature's "salmon factories" around Puget Sound.

The strategy of creating an open-space system along river corridors could provide long-term refuges to anchor regional salmon recovery. Allowing valley-bottom riparian forests to reestablish themselves would lead to the natural restoration of channel processes that create and maintain salmon habitat. This open-space strategy could be implemented gradually through a floodplain buyout program, a ban on development within historically active river corridors, and levee removal. Alternatively, simply stopping the direct and indirect subsidies to floodplain dwellers, such as moneys for levee maintenance and controlling bank erosion, would eventually render continued occupation of these areas unattractive or unfeasible.

Salmon sanctuaries need not be in wilderness. Integrating the remaining non-urban river valleys into a system of salmon sanctuaries connecting Puget Sound to the Cascade Range could be modeled on the San Francisco Bay Area's network of public open spaces. There, over a period of decades, regional special-purpose districts that span several counties gradually accumulated property into a network of publicly accessible open space on the ridges surrounding San Francisco Bay. Creating a network of public open space can not only preserve "wild" areas and connect them in corridors for wildlife but also provides an extensive network of recreational resources such as bike trails and parklands.

Degradation of the Pacific Northwest's rivers and salmon occurred progressively over 150 years and was caused by both deliberate decisions and their inadvertent side effects. River restoration could take even longer, although river rehabilitation programs could accelerate the process of recovery. Over time, rivers will begin to restore themselves if floodplains are reforested and channels are allowed once again to migrate across forested valley bottoms. In the end, the degree to which society is willing to give space back to rivers will define the degree to which rivers can recover. If we don't rehabilitate or restore salmon habitat, more salmon runs will slide toward extinction through our inaction, even if we have not chosen deliberately to sacrifice them for short-term economic gain.

Reclaiming portions of floodplains as more natural river corridors makes sound economic sense. Millions, and over time billions, of dollars could be saved in damage and emergency response alone. Indirect subsidies that allow homes to be developed and businesses to operate in areas prone to flooding would also cease. Furthermore, creating salmon sanctuaries and public open spaces along river corridors would make the surrounding communities more desirable places to live. Enhancing the regional quality of life will become even more important for maintaining regional prosperity in the information age as the population becomes more mobile than ever before.

It should not be controversial that a sustainable fishing industry must be based on the biological realities of the target species. In the case of salmon, the number of fish harvested could be that portion of a run equal to the difference between the number of returning salmon and the number of spawners that had originally produced the returning run. If more than four fish return on average for each spawning pair it could prove sustainable to harvest half of the run, with run size limited by the ability of habitat to sustain salmon and the variability of ocean conditions. Historically, intensive salmon fisheries took 80 to 90 percent of the fish—a harvest rate way beyond what could be sustained biologically. The general principle of 50 percent escapement has worked well for the Alaskan salmon fishery. Taking half of the fish may very well prove sustainable over the long term. Fifty percent or even higher escapement is especially important for the depleted spawning populations that need breathing room now to start rebuilding their numbers.

Native Americans and the Scots conservatively managed their traditional salmon fisheries. Throughout the Pacific Northwest, native peoples allowed large numbers of salmon past their nets and traps to re-seed rivers. Scots followed regulations backed by the authority of a king. Whether by bottom-up customs or top-down edicts, systems of salmon management founded and operated on restraint worked for centuries. Such principles can work again.

Biologist Willis Rich recognized that salmon conservation required reintroduction of restraint as a guiding principle in salmon management. Writing half a century ago, he held that maintenance of Columbia River salmon runs

will mean on the part of all elements in the salmon industry the sacrifice of immediate gain for the benefit of the future. . . . If the courage is lacking now to take the steps necessary to sensible conservation, we shall have the losses and the hardships eventually anyway; with the difference that, if action is delayed, depletion will have progressed further and rehabilitation made just so much more difficult. (Rich 1940, 46)

Today Rich's words seem prophetic.

What steps can be taken now to ensure future healthy fish runs? A logical step would be to ban commercial fishing for at-risk salmon runs altogether for some time. One way to effect this would be to buy out the fishing fleet, the only rapid experiment in adaptive management that could be run on the scale of the entire fishery. But this option is not even discussed in salmon-recovery plans. However, even a total ban on fishing may not stabilize salmon runs if their habitat continues to disappear, or if ocean conditions are unfavorable or deteriorate. Given this reality, it is difficult to ask the fishing industry to sacrifice its present to preserve its future, because it wouldn't work if that is all that is done.

Still, from the perspective of protecting the fishery over the long run, the most sensible approach would be to stop commercial salmon fishing that impacts at-risk wild salmon for five to ten years, several salmon life cycles. When these fisheries reopen, access restrictions could limit harvest to at most half of each run. In addition, allowing salmon to be caught only at the mouth of the river from which they originated would allow stock-specific management plans to be developed, monitored, and if necessary enforced. Not coincidentally, the resulting system would resemble pre-contact Native American salmon management.

A temporary ban on open-water salmon fishing may hit fishing communities hard at first, but such drastic measures may be the only way to prevent rather than simply delay the demise of commercial salmon fishing. Governments have been, and continue to be, more concerned with potential short-run economic dislocations from restrictions aimed at protecting salmon than with preservation of the public resource, salmon themselves. Managers have consistently based allowable catches not on what the fishery could bear over the long run, but on short-term economic interests of fisheries owners and their employees.

Through such actions, governments have put themselves in the position of subsidizing the destruction of a valuable public resource by allowing more people to stay in the fishing industry than the resource itself can support. Eventually some get squeezed out anyway or the industry simply exhausts the supply of fish and collapses altogether.

Salmon are not alone in their predicament. Human actions are transforming Earth into a world to which not only salmon but also many other creatures are not well adapted. We are living through one of the greatest mass extinctions in Earth's history. Millions of years from now, the fossil record of the present age will look a lot like that of the Cretaceous-Tertiary transition, when the dinosaurs and much other life was killed off 65 million years ago. Will salmon be one of the casualties once the dust settles from the Homocene (human) Era extinction now in progress?

We know more about salmon than ever before. But are we using our knowledge of salmon ecology and fluvial geomorphology to figure out how hard we can press the resource before it collapses, perhaps forever? Many technical debates about the details of salmon recovery proposals seem to be animated by this kind of brinkmanship. In truth, the key to restoring salmon is not our knowledge of fish or streams but our ability to manage ourselves.

There is a good chance that we may well lose salmon under recovery plans currently being developed or promoted by politicians and local, state, and federal agencies charged with saving endangered salmon runs in the Northeast and Pacific Northwest. Most of these plans include sensible proposals to reduce actions and practices known to harm salmon. Recovery, however, requires reversing trends, not simply slowing the rate of decline.

As a scientist, I am dismayed by the common tactic of pleading scientific uncertainty as an excuse for continuing to allow people or companies to do things known to harm salmon. We are seldom, if ever, absolutely certain about forecasting the behavior of complex, interacting natural systems, especially for species with complicated life histories. From a public policy perspective, the appropriate question in regard to salmon is whether there is sufficient scientific information upon which informed, responsible management actions can be based—and whether those actions will not only slow but reverse current trends.

There is also a danger in overinterpreting good news. The last several years have seen dramatic increases in salmon returns to many rivers in the Pacific Northwest. Is this dramatic turnaround due to salmon recovery efforts? Is the salmon crisis over? In 2002 fly-fishermen on southern Oregon's Rogue River caught sixty- and seventy-pound chinook, the likes of which hadn't been seen in the river for several decades. It certainly is tempting to credit salmon-recovery and conservation efforts with the recent upswing in salmon returns in the Pacific Northwest. But most of the credit is probably due to increased marine survival resulting from a period of favorable conditions in the oceans. Because the cooler ocean temperatures that support a greater abundance of food for salmon and fewer competitors cycle in and out over decades, it would be foolish to assume that our rivers are restored and that the salmon crisis is over. What will happen when, in a few short years, marine conditions become poor once again for the King of Fish?

The generally favorable ocean conditions for salmon forecast for the first decade of the twenty-first century may well be our last chance to rebuild wild salmon populations across much of the Pacific Northwest. The anticipated higher-than-average marine survival in the next decade can work synergistically with habitat restoration and fishing and hatchery reforms to start rebuilding wild salmon runs before the next downswing in the marine cycle. For we can be sure that marine conditions will reverse, and once again increase the stress on the Pacific Northwest's remaining salmon. What we do in the meantime will prove pivotal. If we fail to take advantage of favorable ocean conditions to rebuild wild salmon runs we may well see at-risk populations disappear the next time ocean conditions turn against them. We need to seize the day and increase wild salmon runs instead of simply harvesting the increase and using the short-term bounty to delay reforms needed for maintaining salmon fisheries over the long term.

But political will is weak. Officials predictably offer up actions designed to avoid controversy rather than address the needs of salmon and protect this vanishing public treasure. Endless calls for more studies, planning, and pilot projects provide a recipe for extinction by default. Regional salmon recovery plans still avoid dealing with some of the most fundamental and most politically charged factors that affect salmon.

Sometimes officials just ignore bad news. In 2001, the Independent Scientific Advisory Board charged with assessing plans to restore wild salmon in the Columbia River Basin concluded that none of the four federal or state plans it reviewed was likely to succeed. A year later, a scientist with the Environmental Protection Agency, Robert Lackey, noted that there had been "almost no reaction from the proponents and managers of the four plans." It is one thing to recognize that public policy decisions involve more than scientific input—scientists have no particular right to set societal objectives. But it is another thing altogether for policy makers to ignore the opinion of scientific panels that their plans, developed to achieve particular policy objectives such as salmon recovery, have little chance for success.

Other times officials help create bad news for salmon. Under President George W. Bush, the U.S. government chose not to defend Endangered Species Act listings of salmon against legal challenges by development and local interests. On the Oregon coast, a local antiregulation group, the Alsea Valley Alliance, claimed that ESA listings of coho salmon inappropriately considered hatchery fish as different from wild fish. The group's position was that since there appeared to be plenty of hatchery-raised fish returning to local streams, why did it matter what was happening to wild stocks? They convinced Federal District Court Judge Michael R. Hogan to revoke the ESA listing for coho in coastal Oregon's Alsea River, and the Bush Administration did not appeal the ruling. Faced with a similar challenge by homebuilders' associations over the adequacy of economic analyses for designating critical habitat for salmon, the government simply chose to suspend the ESA designation while the agency conducted more analyses. These administrative fiats neatly sidestep the issue of how endangered wild salmon are supposed to persist without protection for themselves or their habitat.

Those who assert that it will cost too much to protect salmon often invoke economic arguments and rationalizations. But what does it actually cost not to develop a floodplain, *not* to cut the trees along a stream? Forgoing the opportunity to make money is not the same as losing money. What constitutes an unacceptable economic impact naturally depends both on one's perspective and stake in an issue. Companies and individuals seeking to maximize their profits will, of course, try to protect and expand their ability to make money. Landowners and commercial interests argue that the missed revenue cuts unreasonably into their potential profits. But the public has an equal right to insist upon practices that safeguard salmon and other public resources. Ultimately, society has a responsibility to curtail urges to maximize profits obtained at the expense of the future.

A fundamental problem facing salmon-recovery efforts is that attaining the objective requires taking a new approach to land and river management, one in which our obligation to future generations to be responsible stewards of natural resources is a top priority. Unlike our ancestors, those of us alive today comprise the generations running headlong into the limits of our use of natural systems while observing permanent loss of much of our natural heritage. The bottom line is that people have the freedom to change their behavior, whereas fish do not. If we are to save wild salmon, then some people will lose money or the ability to do things they wanted to do. But we all lose if we lose the salmon.

At the root of the problem is that we generally fail to take into account how landscapes function in the ways we use land. Seeing the dots, we lack mechanisms and policies to connect them. The political time scales of our decision-making processes are poorly matched to the geomorphological time scales that drive landscape change. And our planning and regulatory arenas generally don't account for property wrongs—the downstream impacts of upstream land uses. Instead, society remains fixated on concepts of property rights in which both downstream features and public resources held in common are devalued in favor of a property owner's total freedom of action. The cumulative impacts of private actions on public resources—like salmon—drive home the need to provide economic disincentives for land uses and actions that have a permanent negative ripple effect on their surroundings.

We can have a landscape full of both people and salmon. These goals are not mutually exclusive. But we cannot have both if we are unwilling to fundamentally reexamine how we operate on the landscape and adjust our individual and collective actions so as not to radically degrade the places that salmon live and rely on. If we are serious about wild-salmon recovery, then we need to make tough decisions and changes before salmon are either too far in decline or gone completely. This will not happen without leadership from local, state, and federal governments to identify and aggressively pursue measures that will make the needed changes less disruptive to affected communities. These could include financial assistance and maximization of the economic opportunities created by healthy rivers. We have conclusively proved that we cannot engineer our way around the loss of salmon habitat. Isn't it time to bite the bullet, restore our rivers, and direct more energy, resources, and ingenuity to developing solutions to the economic issues associated with meaningful salmon-recovery actions—solutions well within our grasp if we can muster the political will?

Restoration of Pacific Northwest rivers is not a fanciful daydream. The Puget Sound region still has options for allowing salmon to coexist with a large and growing human population. We still can choose a future with healthy salmon runs and design policies to achieve that vision, but options narrow with each passing year. Whatever the plan, recovery of Puget Sound salmon requires clear vision, strategic thinking, and forceful leadership.

Salmon are a very good, possibly too good, example of the disconnect between wanting to have a plentiful resource but being unwilling to pursue actions that would achieve that goal. No one wants salmon to go extinct, yet we don't seem able to develop and enforce policies to ensure their survival. Policies we do have in place focus on the symptoms and not the causes of salmon declines.

Over the long run, the most vexing problem facing salmon conservation efforts lies in the nature of environmental decision making. In particular, there is a profound mismatch between decisions affecting the places inhabited by salmon and the time it takes for the consequences of those decisions to irreversibly transform the landscape. Put simply, the problem is that little acts gradually pile up into big effects.

David Bella, a professor of engineering at Oregon State University, maintains that we don't need a full scientific understanding to manage a complex system if we use caution as our guiding principle (Bella 2000). Engineers use the concept of a factor of safety—in other words, caution—when designing complex systems. If you really, really want something to work right you don't have a scientist make it; you use an engineer. A good engineer working according to the factor of safety concept will use standards of practice to evaluate what is needed to achieve performance specifications and then will develop a design with a generous margin for error. Engineers study their mistakes and then overdesign their structures. As a result, bridges don't tend to fall down. But when they do, engineers flock from all over to see what happened, to understand the failure, and to learn the lessons for use in future designs.

Not so for river restoration and natural resource managers. They don't look back, they don't undertake retrospective analyses of restoration projects. They move on to the next project and repeat their old mistakes. They don't use factors of safety. They design as close to the edge as they feel they can get away with rather than making sure to stay well back from the edge of uncertainty. Natural-resource managers must learn to assess their past failures, question their assumptions, and work with factors of safety. Even Livingston Stone, the philosophical godfather of hatchery proponents, argued for salmon preserves to safeguard against failure. Without a margin for error, salmon extinction is not only a possibility—it becomes the backup plan.

Professor Bella offers a simple example to explain how societal values and objectives can become disconnected from achieving common, desired goals and lead to outcomes most people actually consider undesirable, perhaps even unacceptable. Imagine that every decision regarding permits for new land development is to be decided from now on by a simple coin toss. Heads allows the project to be built, tails means the project is turned down. One might imagine that this simple lottery-like system gives everyone equal footing and provides a fair (although arbitrary) way to limit the environmental effects of land development. But projects not approved the first time can be proposed again later—a developer can wait and roll again (and again) until Bella's coin toss comes up heads. According to this scenario, which is not far from how landscape change occurs in the real world, it is just a matter of time before everything gets built on.

Add to Bella's coin-toss scenario the reality that, once built, the environmental impacts of many projects cannot be undone, and it becomes clear that is not enough to have good intentions, or even fair laws and plans. Even if Silicon Valley were to be stripped of pavement, there is no topsoil left in which to replant the orchards that once covered the valley. In the case of salmon, each individual project may have little impact, and none may be intended to harm them. But extinction by a thousand cuts, whether deliberate or incidental, is still forever.

Salmon are amazing. Just a few feet long, they travel thousands of miles to complete their life cycle. They can repopulate streams devastated by volcanic eruptions. Given half a chance they can take care of their own existence quite well and expand to fill the available habitat. But for a century and a half we have sustained a pace of landscape-scale changes that salmon have never experienced before except over short time periods and across limited portions of their range. By disturbing everything everywhere all at once, we risk leaving them no sanctuary from which to repopulate depleted rivers and streams.

It is unlikely that we will drive salmon from the face of the earth in my lifetime. At least not until Alaska and Siberia follow the progression of events that over the centuries transformed the rest of Salmon Country. The real issue is how far we will push salmon toward that end, and how many more extinctions of local populations will be allowed in the meantime.

If we can't afford healthy rivers and streams to protect our salmon, then who in the world can? I believe that most people do want salmon in our rivers. Most people tend to want to do the right thing by salmon and are willing to make some sacrifices to prevent the degradation of our environment. But modern salmon management is plagued by short-term thinking that prioritizes vested interests over public interests in setting public policy. This creates the problem that our social and political institutions can lead to outcomes quite different from those which most people desire.

Salmon are like a natural bank account. Generations of Native Americans and Scots lived off of the interest from their accounts. Industrialization of fisheries and transformation of the landscape has eaten deeply into the principal, and depleted the account. As with any bank account, the way to rebuild the balance is to preserve the principal and reinvest the interest—unlike the current management system where fishing intensity rises in good years instead of allowing greater escapement and thus larger spawning runs. Keeping our salmon account solvent over the long run will require returning to the proven practice of only withdrawing the interest.

We borrow salmon from future generations. If it is wrong to destroy something borrowed from a friend—a car, guitar, or whatever—is it not just as wrong to destroy salmon that we borrow from the future? Failure to see the inevitable is a definition of delusion. Sadly, this has been the defining characteristic of salmon management since King George polished off the Thames River salmon.

In the end, the resurrection or destruction of salmon will come down to moral and ethical issues—to value choices that society can make explicitly or continue to make implicitly. Do we want salmon in our rivers? Are we willing to drive species knowingly to extinction, even if only by looking the other way? The solution is not really all that mysterious. We simply cannot keep on doing things the way we've been doing them, or we risk losing the salmon. The choice is ours; the future is not. The sixth H in the salmon story also is ours to choose. Will it be hubris or humility?

RAY TROLL

SOURCES

PREFACE

Rosellini, Hon. Albert D. 1963. "Address of Welcome." In R. S. Crocker and D. Reed, eds., *Report of Second Governor's Conference on Pacific Salmon*, p. 9. Olympia, Wash.: State Printing Plant.

CHAPTER 1: HISTORY, THE FIFTH H

Meeker, E. 1921. *Seventy Years of Progress in Washington*. Tacoma: Allstrum Printing Company.
Montgomery, D. R., S. Bolton, D. B. Booth, and L. Wall, eds. 2003. *Restoration of Puget Sound Rivers*. Seattle and London: University of Washington Press.

CHAPTER 2: SALMON COUNTRY

Abbe, T. B., and D. R. Montgomery. 1996. "Interaction of Large Woody Debris, Channel Hydraulics and Habitat Formation in Large Rivers." *Regulated Rivers: Research & Management* 12: 201–221.
———. 2003. "Patterns and Processes of Wood Debris Accumulation in the Queets River Basin, Washington." *Geomorphology* 51: 81–107.
Collins, B. D., D. R. Montgomery, and A. Haas. 2002. "Historical Changes in the Distribution and Functions of Large Wood in Puget Lowland Rivers." *Canadian Journal of Fisheries and Aquatic Sciences* 59: 66–76.
Montgomery, D. R., J. M. Buffington, P. Peterson, D. Scheutt-Hames, and T. P. Quinn. 1996. "Streambed Scour, Egg Burial Depths and the Influence of

Salmonid Spawning on Bed Surface Mobility and Embryo Survival." *Canadian Journal of Fisheries and Aquatic Sciences* 53: 1061–70.

Montgomery, D. R., G. Pess, E. M. Beamer, and T. P. Quinn. 1999. "Channel Type and Salmonid Spawning Distributions and Abundance." *Canadian Journal of Fisheries and Aquatic Sciences* 56: 377–87.

Winthrop, T. 1863. *The Canoe and the Saddle*. Boston: Ticknor and Fields.

CHAPTER 3: MOUNTAINS OF SALMON

Beechie, T. J., B. D. Collins, and G. R. Pess. 2001. "Holocene and Recent Changes to Fish Habitats in Two Puget Sound Basins." In J. M. Dorava, D. R. Montgomery, B. Palcsak, and F. Fitzpatrick, eds., *Geomorphic Processes and Riverine Habitat*, pp. 37–54. Washington, D.C.: American Geophysical Union.

Buck, R. 1993. *Silver Swimmer: The Struggle for Survival of the Wild Atlantic Salmon*. New York: Lyons & Burford.

Casteel, R. W. 1974. "Use of Pacific Salmon Otoliths for Estimating Fish Size, with a Note on the Size of Late Pleistocene and Pliocene Salmonids." *Northwest Science* 48: 175–79.

Cederholm, C. J., D. B. Houston, D. L. Cole, and W. J. Scarlett. 1989. "Fate of Coho Salmon (*Oncorhynchus Kisutch*) Carcasses in Spawning Streams." *Canadian Journal of Fisheries and Aquatic Sciences* 46: 1347–55.

Devlin, R. H. 1993. "Sequence of Sockeye Salmon Type 1 and 2 Growth Hormone Genes and the Relationship of Rainbow Trout with Atlantic and Pacific Salmon." *Canadian Journal of Fisheries and Aquatic Sciences* 50: 1738–48.

Dimmick, W. W., M. J. Ghedotti, M. J. Grose, A. M. Maglia, D. J. Meinhardt, and D. S. Pennock. 1999. "The Importance of Systematic Biology in Defining Units of Conservation." *Conservation Biology* 13: 653–60.

Frissell, C. A. 1993. "Topology of Extinction and Endangerment of Native Fishes in the Pacific Northwest and California (U.S.A.)." *Conservation Biology* 7: 342–54.

Gross, M. R., R. M. Coleman, and R. M. McDowall. 1988. "Aquatic Productivity and the Evolution of Diadromous Fish Migration." *Science* 239: 1291–93.

Helfield, J. M., and R. J. Naiman. 2001. "Effects of Salmon-Derived Nitrogen on Riparian Forest Growth and Implications for Stream Habitat." *Ecology* 82: 2403–9.

Hendry, A. P., J. K. Wenburg, P. Bentzen, E. C. Volk, and T. P. Quinn. 2000. "Rapid Evolution of Reproductive Isolation in the Wild: Evidence from Introduced Salmon." *Science* 290: 516–18.

Hilderbrand, G. V., S. D. Farley, C. T. Robbins, T. A. Hanley, K. Titus, and C. Serkveen. 1996. "Use of Stable Isotopes to Determine Diets of Living and Extinct Bears. *Canadian Journal of Zoology* 74: 2080–88.

Kendall, A. W., Jr., and, R. J. Behnke. 1983. "Salmonidae: Development and Relationships." *Ontogeny and Systematics of Fishes* (American Society of Ichthyologists and Herpetologists), Special Publication 1, pp. 142–49.

Lucas, R. E. 1986. "Recovery of Game Fish Populations Impacted by the May 18, 1980 Eruption of Mount St. Hellens." In S. A. C. Keller, ed., *Mount St. Helens: Five Years Later,* pp. 276–292. Cheney, Wash.: Eastern Washington University Press.

McKay, S. J., R. H. Devlin, and M. J. Smith. 1996. "Phylogeny of Pacific Salmon and Trout Based on Growth Hormone Type–2 and Mitochondrial NADH Dehydrogenase Subunit 3 DNA Sequences." *Canadian Journal of Fisheries and Aquatic Sciences* 53: 1165–76.

McPail, J. D., 1997. "The Origin and Speciation of *Oncorhynchus* Revisited." In D. J. Stouder, P. A. Bisson, and R. J. Naiman, eds., *Pacific Salmon and Their Ecosystems,* pp. 29–38. New York: Chapman & Hall.

Milner, A. M. 1987. "Colonization and Ecological Development of New Streams in Glacier Bay National Park, Alaska." *Freshwater Biology* 18: 53–70.

Montgomery, D. R. 2000. "Coevolution of the Pacific Salmon and Pacific Rim Topography." *Geology* 28: 1107–10.

Nehlsen, W., J. E. Williams, and J. A. Lichatowich. 1991. "Pacific Salmon at the Crossroads: Stocks at Risk from California, Oregon, Idaho, and Washington." *Fisheries* 16(2): 4–21.

Porter, S. C., and T. W. Swanson. 1998. "Radiocarbon Age Constraints on Rates of Advance and Retreat of the Puget Lobe of the Cordilleran Ice Sheet During the Last Glaciation." *Quaternary Research* 50: 205–13.

Quinn, T., 2003, personal communication.

Smith, G. R. 1981. "Late Cenozoic Freshwater Fishes of North America." *Annual Reviews of Ecology and Systematics* 12: 163–93.

Stearley, R. F. 1992. "Historical Ecology of Salmoninae, with Special Reference to *Oncorhynchus.*" In R. L. Mayden, ed., *Systematics, Historical Ecology, and North American Freshwater Fishes*. Palo Alto, Calif.: Stanford University Press, pp. 622–58.

Stearley, R. F., and G. R. Smith. 1993. "Phylogeny of the Pacific Trouts and Salmons (*Oncorhynchus*) and Genera of the Family Salmonidae." *American Fisheries Society Transactions* 122: 1–33.

Thomas, W. K., R. E. Withler, and A. T. Beckenback. 1986. "Mitochondrial DNA Analysis of Pacific Salmonid Evolution." *Canadian Journal of Zoology* 64: 1058–64.

Wilson, M. V. H., and R. R. G. Williams. 1992. "Phylogentic, Biogeographic, and Ecological Significance of Early Fossil Records of North American Freshwater Teleostean Fishes." In R. L. Mayden, ed., *Systematics, Historical Ecology, and North American Freshwater Fishes*, pp. 224–244. Palo Alto, Calif.: Stanford University Press.

CHAPTER 4: SALMON PEOPLE

American Friends Service Committee. 1970. *Uncommon Controversy: Fishing Rights of the Muckleshoot, Puyallup, and Nisqually Indians*. Seattle and London: University of Washington Press.

Boxberger, D. L. 2000. *To Fish in Common: The Ethnohistory of Lummi Indian Salmon Fishing*. Seattle and London: University of Washington Press.

Boyd, R. 1999. *The Coming of the Spirit of Pestilence: Introduced Infectious Diseases and Population Decline Among Northwest Coast Indians, 1774–1874*. Vancouver and Toronto: University of British Columbia Press, and Seattle and London: University of Washington Press.

Brown, W. C. 1961. *The Indian Side of the Story*. Spokane, Wash.: C. W. Hill Printing Co.

Butler, V. L. 1993. Natural Versus Cultural Salmonic Remains: Origin of the Dalles Roadcut Bones, Columbia River, Oregon, U.S.A. *Journal of Archaeological Science* 20:1–24.

Catlin, G. [1861/1866] 1959. *Episodes from "Life Among the Indians" and "Last Rambles" with 152 Scenes and Portraits by the Artist*. Norman, Okla.: University of Oklahoma Press.

Cone, J., and S. Ridlington, eds. 1996. *The Northwest Salmon Crisis: A Documentary History*. Corvallis: Oregon State University Press.

DeVoto, B., ed. 1953. *The Journals of Lewis and Clark*. Boston: Houghton Mifflin.

Finney, B. P., I. Gregory-Eaves, M. S. V. Douglas, and J. P. Smol. 2002. "Fisheries Productivity in the Northeastern Pacific Ocean Over the Past 2,200 Years." *Nature* 416: 729–33.

Hebda, R., and S. G. Frederick. 1990. "History of Marine Resources of the Northeast Pacific Since the Last Glaciation." *Transactions of the Royal Society of Canada* (series 6) 1: 319–42.

Mantua, N. J., S. R. Hare, Y. Qhang, J. M. Wallace, and R. C. Francis. 1997. "A Pacific Interdecadal Climate Oscillation with Impacts on Salmon Production." *Bulletin of the American Meteorological Society* 78: 1069–79.

McCoy, S. 1991. "Ending Washington State's Long 'Fish War.'" *High Country News*, April 22, 1991, p. 19–21.

Meeker, E. 1905. *Pioneer Reminiscences of Puget Sound and the Tragedy of Leschi*. Seattle: Lowman & Hanford.

Minot, G., ed. 1855. *The Statutes at Large and Treaties of the United States of America. From December 1, 1851, to March 3, 1855*. Vol. X. Boston: Little, Brown and Company.

Pess, G. R., D. R. Montgomery, R. E. Bilby, A. E. Steel, B. E. Feist, and H. M.Greenberg. 2002. "Landscape Characteristics, Land Use, and Coho Salmon (*Oncorhynchus kisutch*) Abundance, Snohomish River, Washington State, USA." *Canadian Journal of Fisheries and Aquatic Sciences* 59: 613–23.

Puget Sound Gillnetters v. United States District Court. 1978. 573 F.2d 112 3 (9th Cir.).

Swezey, S. L., and R. F. Heizer. 1993. "Ritual Management of Salmonid Fish Resources in California." In T. C. Blackburn and K. Anderson, eds., *Before the Wilderness: Environmental Management by Native Californians*, pp. 299–327. Ballena Press Anthropological Paper No. 40. Menlo Park, Calif.: Ballena Press.

Swindell, E. G. 1942. *Report on Source, Nature, and Extent of the Fishing, Hunting, and Miscellaneous Related Rights of Certain Indian Tribes in Oregon and Washington*. Los Angeles: U.S. Department of the Interior, Office of Indian Affairs, Division of Forestry and Grazing.

United States v. Winans. 1905. 198 U.S. 371.

United States v. State of Washington. 1980. 506 F. Supp. 187 (W. D. Wash.).

Washington State Department of Fisheries. 1959. *1959 Annual Report*. Olympia: Department of Fisheries.

———. 1961. *1961 Annual Report*. Olympia: Department of Fisheries.

CHAPTER 5: OLD WORLD SALMON

Anne Regina (Queen Anne). 1712. *An Act for the better Preservation and Improvement of the Fishery within the River of Thames, and for Regulating and Governing the Company of Fishermen of the said River*. Printed by John Baskett, and the Assigns of Thomas Newcomb, and Henry Hill, deceased; Printers to the Queens most Excellent Majesty, London.

Aston, M., ed. 1988. *Medieval Fish, Fisheries and Fishponds in England*. Oxford: British Archeological Reports.

Brown, W. 1862. *The Natural History of the Salmon, as Ascertained by the Recent Experiments in the Artificial Spawning and Hatching of the Ova and Rearing of the Fry, at Stormontfield, on the Tay*. Glasgow: Thomas Murray & Son.

Cornish, J. 1824. *A View of the Present State of the Salmon Fisheries, and of the Statute Laws by Which They are Regulated*. London: Longman, Hurst, Rees, Orme, Brown, & Green.

Edinburgh Review. 1851. "The Salmon Fisheries." *Edinburgh Review* 93: 340–69.

———. 1873. "The English Salmon Fisheries." *Edinburgh Review* 137: 153–82.

Evelyn, J. 1679. *Sylva, Or a Discourse of Forest-Trees, and the Propagation of Timber in His Majesty's Dominions*. London: John Martyn.

Fraser, A. 1833. *Natural History of the Salmon, Herrings, Cod, Ling, &c.; With a Short Account of Greenland, its Inhabitants, Land and Sea Animals, and the Different Tribes of Fishes Found on the Coast*. Inverness, Scotland: R. Carruthers.

Gudjónsson, T. 1988. "Exploitation of Atlantic Salmon in Iceland." In D. Mills and D. Piggins, eds., *Atlantic Salmon: Planning for the Future*, Proceedings of the Third International Atlantic Salmon Symposium, pp. 162–177. London and Sydney: Croom Helm Ltd.

Harwood, K., and A. G. Brown. 1993. "Fluvial Processes in a Forested Anastomosing River: Flood Partitioning and Changing Flow Patterns." *Earth Surface Processes and Landforms* 18: 741–48.

Levinge, R. 1980. "A General Review of the State of the Salmon Fisheries of the North Atlantic." In A. E. J. Went, ed., *Atlantic Salmon: Its Future*, Proceedings of the Second International Atlantic Salmon Symposium, pp. 6–17. Farnham, England: Fishing News Books Ltd.

Netboy, A. 1968. *The Atlantic Salmon: A Vanishing Species?* Boston: Houghton Mifflin.

———. 1974. *The Salmon: Their Fight for Survival*. Boston: Houghton Mifflin.

Petts, G. E., H. Möller, and A. L. Roux, eds. 1989. *Historical Change of Large Alluvial Rivers: Western Europe*. Chichester, England: John Wiley.

Russel, A. 1864. *The Salmon*. Edinburgh: Edmonston & Douglas.

Quarterly Review. 1828. "Salmonia, or Days of Fly-fishing." *The Quarterly Review* 38: 503–35.

———. 1852. "Irish Salmon." *The Quarterly Review* 91: 252–379.

———. 1863. "The Salmon Question." *The Quarterly Review* 113: 388–422.

World Wildlife Fund. 2001. *The Status of Wild Atlantic Salmon: A River by River Assessment*. Washington, D.C.: World Wildlife Fund, U.S. Marine Conservation Program.

Young, A. 1854. *The Natural History and Habits of the Salmon; with Reasons for the Decline of the Fisheries*. London: Longman, Brown, Green, & Longmans.

CHAPTER 6: NEW WORLD SALMON

Bartram, J. 1751. *Observations on the Inhabitants, Climate, Soil, Rivers, Productions, Animals, and other matters worthy of Notice. Made by Mr. John Bartram, In his Travels from Pensilvania to Onondago, Oswego and the Lake Ontario, in Canada.* London: J. Whiston & B. White.

Buck, R. 1993. *Silver Swimmer: The Struggle for Survival of the Wild Atlantic Salmon.* New York: Lyons & Burford.

Commissioners of Fisheries of the State of Maine. 1869. *Reports of the Commissioners of Fisheries of the State of Maine for the Years 1867 and 1868.* 1869. Augusta, Me.: Owen & Nash, Printers to the State.

Congressional Record. 1971. "Statement by Richard A. Buck." *Congressional Record* 117, part 13: 16,561.

Cronon, W. 1983. *Changes in the Land: Indians, Colonists, and the Ecology of New England.* New York: Hill & Wang.

Denys, N. 1908. *The Description and Natural History of the Coasts of North America (Acadia).* Trans. and ed. W. F. Ganong. Publications of the Champlain Society no. 2. Toronto: Champlain Society.

Dunfield, R. W. 1985. *The Atlantic Salmon in the History of North America.* Canadian Special Publication of Fisheries and Aquatic Sciences no. 80. Ottowa: Department of Fisheries and Oceans.

Goode, G. B. 1887. *The Fisheries and Fishery Industries of the United States.* Section 5, Vol. 1. Washington, D.C.: Government Printing Office.

Hardy, C. 1855. *Sporting Adventures in the New World; or Days and Nights of Moose-Hunting in the Pine Forests of Acadia.* Volume 2. London: Hurst & Blackett.

Juet, R. 1909. "From 'The third voyage of Master Henry Hudson,' by Robert Juet, 1610." In J. F. Jameson, ed., *Narratives of New Netherland, 1609–1664,* pp. 16–28. New York: Charles Scribner's Sons.

Kendall, W. C. 1935. *The Fishes of New England. The Salmon Family. Part 2, "The Salmons."* Memoirs of the Boston Society of Natural History: Monographs on the Natural History of New England. Boston: Boston Society of Natural History.

Lee, P. 1996. *Home Pool: The Fight to Save the Atlantic Salmon.* Fredericton, N.B.: Goose Lane Editions and New Brunswick Publishing.

McGregor, J. 1828. *Historical and Descriptive Sketches of the Maritime Colonies of British America.* London: Longman, Rees, Orme, Brown, & Green.

Marsh, G. P. 1864. *Man and Nature; or, Physical Geography as Modified by Human Action.* New York: Charles Scribner.

National Research Council. 2002. *Genetic Status of Atlantic Salmon in Maine: Interim Report from the Committee on Atlantic Salmon in Maine*. Washington, D.C.: National Academy Press.

Netboy, A. 1968. *The Atlantic Salmon: A Vanishing Species?* Boston: Houghton Mifflin.

Pearson, J. C. 1972. *The Fish and Fisheries of Colonial North America. A Documentary History of Fishing Resources of the United States and Canada.* Part 2, "The New England States." Washington, D.C.: U.S. Fish and Wildlife Service, Department of the Interior.

Smith, Capt. John. [1616] 1865. *A Description of New England; or, Observations and Discoveries in the North of America in the Year of Our Lord 1614. With the Success of Six Ships that Went the Next Year, 1615.* London; reprint, Boston: William Veazie.

Smith, J. V. C. 1843. *Natural History of the Fishes of Massachusetts, Embracing a Practical Essay on Angling*. Boston: William D. Ticknor.

Stolte, L. 1981. *The Forgotten Salmon of the Merrimack*. U.S. Department of the Interior, Northeast Region. Washington, D.C.: Government Printing Office.

Thoreau, H. D. 1849. *A Week on the Concord and Merrimack Rivers*. Boston and Cambridge: James Munroe.

Wood, W. 1634. *New Englands Prospect. A True, Lively, and Experimentall Description of that Part of America, Commonly Called New England: Discovering the State of that Countrie, Both as it Stands to Our New-Come English Planters; and to the old Native Inhabitants*. London: John Bellamie.

Chapter 7: Western Salmon Rush

Bancroft, H. H. 1888. *History of Oregon. Volume 2, 1848–1888. The Works of Hubert Howe Bancroft* 30. San Francisco: The History Company.

Brice, J. J. 1895. "Establishment of Stations for the Propagation of Salmon on the Pacific Coast." In U.S. Commission of Fish and Fisheries, *Report of the Commissioner for the Year Ending June 30, 1893*, pp. 387–392. Washington, D.C.: Government Printing Office.

California Division of Fish and Game. 1937. *Thirty-fourth Biennial Report for 1934–1936*. Sacramento: California State Printing Office.

Cobb, J. N. 1930. "Pacific Salmon Fisheries." In *Report of the Commissioner of Fisheries for 1930*, pp. 409–704. Washington, D.C.: U.S. Department of Commerce, Bureau of Fisheries.

Freeman, M. 1952. "The Privilege of Publishing." *Pacific Fisherman* 50 (August): 4.

Gregory, H. E., and K. Barnes. 1939. *North Pacific Fisheries, with Special Reference to Alaska Salmon*. Studies of the Pacific no. 3. San Francisco: American Council, Institute of Pacific Relations.

Hume, R. D. 1893. *Salmon of the Pacific Coast*. San Francisco: Schmidt Label & Lithographic.

Kutchin, H. M. 1898. *Report on the Salmon Fisheries of Alaska, 1897*. U.S. Treasury Department, Division of Special Agents, Document Number 2010, Washington, D.C.: Government Printing Office.

McDonald, M. 1894. "The Salmon Fisheries of the Columbia River Basin." In *Report of the Commissioner of Fish and Fisheries on Investigations in the Columbia River Basin in Regard to the Salmon Fisheries*, pp. 3–18. Washington, D.C.: Government Printing Office.

McGuire, H. D. 1894. *First and Second Annual Reports of the Fish and Game Protector to the Governor, 1893–94*. Salem, Ore.: Frank L. Baker, State Printer.

———. 1896. *Third and Fourth Annual Reports of the Fish and Game Protector of the State of Oregon, 1895–96*. Salem, Ore.: W. H. Leeds, State Printer.

Meeker, E. 1870 *Washington Territory West of the Cascade Mountains, Containing a Description of Puget Sound, and Rivers Emptying into It*. Olympia, Washington Territory: Transcript Office.

Meeker, E. 1921. *Seventy Years of Progress in Washington*. Tacoma: Allstrum Printing Company.

Murray, J. 1898. "Report of Special Agent Murray on the Salmon Fisheries in Alaska." In *Seal and Salmon Fisheries and General Resources of Alaska*. Volume 2, pp. 404–418 (55th Congress, 1st session, House of Representatives, document no. 92, part 2). Washington, D.C.: Government Printing Office.

Newell, D. 1989. *The Development of the Pacific Salmon-Canning Industry: A Grown Man's Game*. Montreal and Kingston: McGill-Queens University Press.

Nobbs, P. E. 1949. *The Miramichi Fisheries*. Document no. 9. Montreal: Atlantic Salmon Association.

Pess, G., D. R. Montgomery, T. J. Beechie, and L. Holsinger. 2003. "Anthropogenic Alterations to the Biogeography of Puget Sound Salmon." In D. R. Montgomery, S. Bolton, D. B. Booth, and L. Wall, eds., *Restoration of Puget Sound Rivers*, pp. 129–54. Seattle and London: University of Washington Press.

Rich, W. H. 1940. "The Future of the Columbia River Salmon Fisheries." *Stanford Ichthyological Bulletin* 2(2): 37, 46.

Smith, C. L. 1979. *Salmon Fishers of the Columbia.* Corvallis: Oregon State University Press.

Stone, L. 1885. "Explorations on the Columbia River from the Head of Clarkeís Fork to the Pacific Ocean, Made in the Summer of 1883, with Reference to the Selection of a Suitable Place for Establishing a Salmon-Breeding Station." In United States Commission of Fish and Fisheries, *Report of the Commissioner for 1883*, pp. 237–55. Washington, D.C.: Government Printing Office.

Suckley, G. 1860. "Report upon the Fishes Collected on the Survey." In *Reports of Explorations and Surveys to Ascertain the Most Practicable and Economical Route for a Railroad from the Mississippi River to the Pacific Ocean.* Volume 12, book 2, pp. 307–99 (36th Congress, 1st session, House of Representatives, executive document no. 56). Washington, D.C.: Government Printing Office.

———. 1861. "Notices of Certain New Species of North American Salmonidae from the North-West Coast of America." *Annals of the Lyceum of Natural History* 7: 306–13.

Swan, J. G. 1857. *The Northwest Coast; Or, Three Years' Residence in Washington Territory.* New York: Harper & Brothers.

van Dusen, H. G. 1903. *Annual Reports of the Department of Fisheries of the State of Oregon to the Legislative Assembly, Twenty-Second Regular Session.* Salem, Ore.: W. H. Leeds, State Printer.

Wendler, H. O. 1966. "Regulation of Commercial Fishing Gear and Seasons on the Columbia River from 1859 to 1963." *Fisheries Research Papers of the Washington Department of Fisheries* 2, no. 4: 19–31.

Wilkes, C. 1844. *Narrative of the United States Exploring Expedition During the Years 1838, 1839, 1940, 1841, 1842.* Volume 4. Philadelphia: C. Sherman.

CHAPTER 8: BETTER THAN NATURAL

Baird, S. 1875. "The Salmon Fisheries of Oregon." *Oregonian*, March 3, 1875.

Brown, W. 1862. *The Natural History of the Salmon, As Ascertained by the Recent Experiments in the Artificial Spawning and Hatching of the Ova and Rearing of the Fry, at Stormontfield, on the Tay.* Glasgow: Thomas Murray & Son.

Cobb, J. N. 1930. "Pacific Salmon Fisheries." In *Report of the Commissioner of Fisheries for 1930*, pp. 409–704. Washington, D.C.: U.S. Dept. of Commerce, Bureau of Fisheries.

Cone, J. 1995. *A Common Fate: Endangered Salmon and the People of the Pacific Northwest.* New York: Henry Holt.

Cone, J., and S. Ridlington, eds. 1996. *The Northwest Salmon Crisis: A Documentary History*. Corvallis: Oregon State University Press.

Darwin, L. H. 1916. *The Fisheries of the State of Washington*. Olympia: Department of State, State Bureau of Statistics and Immigration.

Edinburgh Review. 1873. "The English Salmon Fisheries." *Edinburgh Review* 137: 153–82.

Garlick, T. 1880. *A Treatise on the Artificial Propagation of Fish, with Description and Habits of Such Kinds as are Suitable for Domestic Fish-Culture*. Cleveland, Ohio: J. B. Savage.

Hilborn, R. 1992. "Hatcheries and the Future of Salmon in the Northwest." *Fisheries* 17(1): 5–8.

Jordan, D. S., and B. W. Everman. 1902. *American Food and Game Fishes: A Popular Account of all the species Found in America North of the Equator with Keys for Ready Identification, Life Histories and Methods of Capture*. New York: Doubleday.

Lee, P. 1996. *Home Pool: The Fight to Save the Atlantic Salmon*. Fredericton, N.B.: Goose Lane Editions and New Brunswick Publishing.

McDonald, J. 1981. "The Stock Concept and Its Application to British Columbia Salmon Fisheries." *Canadian Journal of Fisheries and Aquatic Sciences* 38: 1657–64.

McDonald, M. 1894. "The Salmon Fisheries of the Columbia River Basin." In *Report of the Commissioner of Fish and Fisheries on Investigations in the Columbia River Basin in Regard to the Salmon Fisheries*, pp. 3–18. Washington, D.C.: Government Printing Office.

McLeod, K. 1959. "History of Fish Farming." In K. McLeod, ed., *Fisheries: Fish Farming, Fisheries Management*, pp. 11–22. Olympia: Washington State Department of Fisheries.

Miller, W. H. 1990. *Analysis of Salmon and Steelhead Supplementation*. Report to Bonneville Power Administration (BPA Report DOE/BP–92663–1). Portland, Ore.: U.S. Fish and Wildlife Service.

Moring, J. R. 2000. "The Creation of the First Public Salmon Hatchery in the United States." *Fisheries* 25(7): 6–11.

National Research Council. 2002. *Genetic Status of Atlantic Salmon in Maine: Interim Report from the Committee on Atlantic Salmon in Maine*. Washington, D.C.: National Academy Press.

Northwest Power Planning Council. 1998. Scientific Review Team, Independent Scientific Advisory Board. *Review of Artificial Production of Anadromous and Resident Fish in the Columbia River Basin*. Document no. 98–33. Portland, Ore.: Northwest Power Planning Council, Program Evaluation and Analysis Section.

Rich, W. H. 1922. Early History and Seaward Migration of Chinook Salmon in the Columbia and Sacramento Rivers. *Bulletin of the Bureau of Fisheries* 37: 1–74.

———. 1939. "Local Populations and Migration in Relation to the Conservation of Pacific Salmon in the Western States and Alaska." In F. R. Moulton, ed., *The Migration and Conservation of Salmon*, pp. 45–50. American Association for the Advancement of Science Publication no. 8. Lancaster, Pa.: Science Press.

———. 1941. "The Present State of the Columbia River Salmon Resources." *Proceedings of the Sixth Pacific Science Congress* 3 (Berkeley, Calif.): 425–30.

Shanks, W. F. G. 1868. Fish-culture in America. *Harper's New Monthly Magazine* 37: 721–39.

Smith, C. L. 1979. *Salmon Fishers of the Columbia.* Corvallis: Oregon State University Press.

Stone, L. 1885. "Explorations on the Columbia River from the Head of Clarke's Fork to the Pacific Ocean, Made in the Summer of 1883, with Reference to the Selection of a Suitable Place for Establishing a Salmon-Breeding Station." In *Report of the Commissioner for 1883*, pp. 237–55, prepared for United States Commission of Fish and Fisheries. Washington, D.C.: Government Printing Office.

———. 1896. "The Artificial Propagation of Salmon on the Pacific Coast of the United States, with Notes on the Natural History of the Quinnat Salmon." *Bulletin of the United States Fish Commission* 16: 203–35.

U.S. Commission of Fish and Fisheries. 1903. *Artificial Propagation of the Salmons of the Pacific Coast.* Washington, D.C.: Government Printing Office.

U.S. Commissioner of Fisheries. 1937. Report of Commissioner of Fisheries on Bonneville Dam and Protection of Columbia River Fisheries. Report to U.S. Senate (document 87, 75th Congress, 1st session). Washington, D.C.: Government Printing Office.

Ward, H. B. 1939. "Factors Controlling Salmon Migration." In F. R. Moulton, ed., *The Migration and Conservation of Salmon*. American Association for the Advancement of Science Publication no. 8. Lancaster, Pa.: Science Press.

Washington State Department of Fisheries. 1968. *1968 Annual Report.* Olympia: Washington State Department of Fisheries.

World Wildlife Fund. 2001. *The Status of Wild Atlantic Salmon: A River by River Assessment.* Washington, D.C.: World Wildlife Fund, U.S. Marine Conservation Program.

Young, A. 1854. *The Natural History and Habits of the Salmon; with Reasons for the Decline of the Fisheries*. London: Longman, Brown, Green, & Longmans.

CHAPTER 9: POWER FOR THE PEOPLE

Anderson, A. 1950. "Shall We Have Salmon, or Dams, or Both?" In E. M. Quee, ed., *Transactions of the Fifteenth North American Wildlife Conference*, pp. 449–54. Washington, D.C.: Wildlife Management Institute.

Bell, F. T. 1937. "Guarding the Columbia's Silver Horde." *Nature Magazine* 29 (January): 43–47.

Bonneville Power Administration. N.d. *Columbia River Power for the People: A History of Policies of the Bonneville Power Administration*. Publication no. DOE-BP-7. Portland, Ore.: U.S. Department of Energy, Bonneville Power Administration.

Cone, J., and S. Ridlington, eds. 1996. *The Northwest Salmon Crisis: A Documentary History*. Corvallis: Oregon State University Press.

Douglas, D. [1823–27] 1959. *Journal Kept by David Douglas During his Travels in North America, 1823–1827*. New York: Antiquarian Press.

Everman, B. W., and S. E. Meek. 1898. "A Report Upon Salmon Investigations in the Columbia River Basin and Elsewhere on the Pacific Coast in 1896." *Bulletin of the United States Fish Commission* 17: 15–84.

Idaho Department of Fish and Game v. National Marine Fisheries Service. 1994. 850 F. Supp. 886 (D.Or.).

Johnson, J. H. 1960. "Sonic Tracking of Adult Salmon at Bonneville Dam, 1957." *Fishery Bulletin of the Fish and Wildlife Service* 60: 471–85.

Netboy, A. 1958. *Salmon of the Pacific Northwest: Fish vs. Dams*. Portland, Ore.: Binfords & Mort.

Northwest Resource Information Center v. Northwest Power Planning Council. 1994. 35 F.3d 1371 (9th Cir.).

Pacific Fisherman. 1903. "Dams Jeopardize Fish Industry." *Pacific Fisherman* 1(3): 5.

Pollock, C. 1932. Fortieth and Forty-First Annual Reports of the State Department of Fish and Game, Division of Fisheries. Olympia, Wash.: State of Washington, Department of Fisheries and Game.

Udall v. Federal Power Commission. 1967. 387 U.S. 428.

U.S. Commissioner of Fisheries. 1937. *Report of Commissioner of Fisheries, on Bonneville Dam and Protection of Columbia River Fisheries*. Report to U.S. Senate (document 87, 75th Congress, 1st session). Washington, D.C.: Government Printing Office.

U.S. Fish and Wildlife Service. 1958. *Fishery Development Program of the Columbia River. Report of Activities to June 30, 1958.* Portland, Ore.: Department of the Interior, Fish and Wildlife Service, Bureau of Commercial Fisheries.

U.S. House of Representatives. 1952. *Hearings Before the Subcommittee of the Committee on Appropriations, Civil Functions, Department of the Army, Appropriations for 1953* (82nd Congress, 2nd Session). Washington, D.C.: Government Printing Office.

Ward, H. B. 1938. *Placer Mining on the Rogue River, Oregon, in Its Relation to the Fish and Fishing in that Stream.* Oregon Department of Geology and Mineral Industries, bulletin no. 10. Portland, Ore.: State of Oregon, Department of Geology and Mineral Industries.

Winn, D., C. D. Shoemaker, E. Seaborg, J. N. Cobb, E. D. Clark, and M. Freeman. 1924. *Save the Columbia River Salmon; Brief Submitted to Federal Power Commission in Opposition to Application of Washington Irrigation and Development Company for a License to Construct a Dam Across the Columbia River at Priest Rapids, Washington.* Olympia, Wash.: Frank M. Lamborn.

CHAPTER 10: RIVERS OF CHANGE

Beckham, D. 1990. *Swift Flows the River.* Coos Bay, Ore.: Arago Books.

Beechie, T., E. Beamer, and L. Wasserman. 1994. "Estimating Coho Salmon Rearing Habitat and Smolt Production Losses in a Large River Basin, and Implications for Habitat Restoration." *North American Journal of Fisheries Management* 14: 797–811.

Collins, B. D., and D. R. Montgomery. 2002. "Forest Development, Wood Jams, and Restoration of Floodplain Rivers in the Puget Lowland, Washington." *Restoration Ecology* 10: 237–47.

Collins, B. D., D. R. Montgomery, and A. J. Sheikh. 2003. Reconstructing the Historical Riverine Landscape of the Puget Lowland. In D. R. Montgomery, S. Bolton, D. B. Booth, and L. Wall, eds., *Restoration of Puget Sound Rivers*, pp. 79–128. Seattle and London: University of Washington Press.

Froehlich, H. A. 1973. "Natural and Man-Caused Slash in Headwater Streams." *Loggers Handbook* 33, pp. 15–17, 66–86. Portland, Ore.: Pacific Logging Congress.

Gharrett, J. T., and J. I. Hodges. 1950. "Salmon Fisheries of the Coastal Rivers of Oregon South of the Columbia." Contribution no. 13. Portland, Ore.: Oregon Fish Commission.

Graf, W. L. 1999. "Dam Nation: A Geographic Census of American Dams and Their Large-Scale Hydrologic Impacts." *Water Resources Research* 35: 1305–11.

Hart, J. L. 1950. "Fishery Problems in British Columbia." In E.M. Quee, ed., *Transactions of the Fifteenth North American Wildlife Conference*, pp. 421–26. Washington, D.C.: Wildlife Management Institute.

Huntington, C., W. Nehlsen, and J. Bowers. 1996. "A Survey of Healthy Native Stocks of Anadromous Salmonids in the Pacific Northwest and California." *Fisheries* 21(3): 6–14.

An Illustrated History of Skagit and Snohomish Counties. 1906. Chicago, Ill.: Interstate Publishing Company.

Locke, S. B. 1929. *Whitefish, Grayling, Trout, and Salmon of the Intermountain Region*. U. S. Department of Commerce, Bureau of Fisheries, Document No. 1062. Washington, D.C.: U.S. Government Printing Office.

McLeod, A. R. [1826] 1961. "Journal of a Trapping Expedition along the Coast South of the Columbia in charge of A. R. McLeod C. T., Summer 1826." In K. G. Davies, ed., *Peter Skene Ogden's Snake Country Journal, 1826–27*. Publication no. 23. London: Hudson's Bay Record Society.

National Research Council. 1992. *Restoration of Aquatic Ecosystems*. Washington, D.C.: National Academy Press.

Pess, G. R., D. R. Montgomery, R. E. Bilby, A. E. Steel, B. E. Feist, and H. M. Greenberg. 2002. "Landscape Characteristics, Land Use, and Coho Salmon (*Oncorhynchus kisutch*) Abundance, Snohomish River, Washington State, U.S.A." *Canadian Journal of Fisheries and Aquatic Sciences* 59: 613–23.

Ruffner, E. H. 1886. *The Practice of the Improvement of the Non-Tidal Rivers of the United States, with an Examination of the Results Thereof*. New York: John Wiley.

Sedell, J. R., P. A. Bisson, J. A. June, and R. W. Speaker. 1982. "Ecology and Habitat Requirements of Fish Populations in South Fork Hoh River, Olympic National Park." In *Ecological Research in National Parks of the Pacific Northwest; Proceedings of the Second Conference on Scientific Research in the National Parks*. Corvallis: Oregon State University Forest Research Laboratory.

Sedell, J. R., J. E. Yuska, and R. W. Speaker. 1984. "Habitats and Salmonid Distribution in Pristine, Sediment-Rich River Valley Systems: S. Fork Hoh and Queets River, Olympic National Park." In *Fish and Wildlife Relationships in Old-Growth Forests*, pp. 33- 46. Morehead City, North Carolina: American Institute of Fishery Biologists.

Sedell, J. R., and J. L. Froggatt. 1984. "Importance of Streamside Forests to Large Rivers: The Isolation of the Willamette River, Oregon, U.S.A., from

Its Floodplain By Snagging and Streamside Forest Removal." *Verhandlungen—Internationale Vereinigung für Theoretische und Angewandte Limnologie* 22: 1828–34.

Sedell, J. R., F. N. Leone, and W. S. Duval. 1991. "Water Transportation and Storage of Logs." In *Influences of Forest and Rangeland Management on Salmonid Fishes and Their Habitats*, pp. 325–68. Special Publication 19. Bethesda: American Fisheries Society.

Shukovsky, P. 2001. "Hostage to a River's Ups and Downs." *Seattle Post-Intelligencer*, December 22, pp. A1, A12.

Stover, S. C., and D. R. Montgomery. 2001. "Channel Change and Flooding, Skokomish River, Washington." *Journal of Hydrology* 243: 272–86.

Thomas, B. P. 1964. "King County, Washington, Comprehensive Plan for Flood Control." Report prepared by Bertram P. Thomas, consulting engineer, for the board of King County Commissioners.

Triska, F. J. 1984. "Role of Wood Debris in Modifying Channel Geomorphology and Riparian Areas of a Large Lowland River Under Pristine Conditions: A Historical Case Study." *Verhandlungen-Internationale Vereinigung für Theoretische und Angewandte Limnologie* 22: 1876–92.

U.S. House of Representatives. 1875. "Improvement of the Willamette River above Oregon City, Oregon." In *Report of the Secretary of War* (43rd Congress, 1st session, executive document 1–2), appendix GG2, pp. 760–772. Washington, D.C.: Government Printing Office.

———. 1881a. "Examination of the Stillaguamish River, Washington Territory." In *Report of the Chief of Engineers, U.S. Army* (47th Congress, 1st session, executive document 1–5), appendix OO11, pp. 2608–11. Washington, D.C.: Government Printing Office.

———. 1881b. "Improvement of Skagit River, Washington Territory." In *Report of the Chief of Engineers, U.S. Army* (47th Congress, 1st session, executive document 1–5), appendix OO10, pp. 2603–8. Washington, D.C.: Government Printing Office.

Washington State Department of Fisheries. 1951. *The Salmon Crisis*. Seattle, Wash.: Department of Fisheries.

———. 1965. *1965 Annual Report*. Olympia, Wash.: Department of Fisheries.

———. 1966. *1966 Annual Report*, pp. 117–223. Olympia, Wash.: Department of Fisheries.

———. 1967. *A Positive Program to Maintain and Enhance Our Fisheries: The Department of Fisheries' 10-Year Plan to Maintain, Enhance and Increase the Foodfish Fisheries of Washington State*. Olympia, Wash.: Department of Fisheries.

Wendler, H. O., and G. Deschamps. 1955. "Logging Dams on Coastal Washington Streams." *Washington Department of Fisheries, Fisheries Research Papers* 1: 27–38.

Wilkes, C. 1844. *Narrative of the United States Exploring Expedition During the Years 1838, 1839, 1840, 1841, 1842*. Volume 5. Philadelphia: C. Sherman.

CHAPTER 11: THE SIXTH H

Bella, D. 2000. Personal communication.

Gresh, T., J. Lichatowich, and P. Schoonmaker. 2000. "An Estimation of Historic and Current Levels of Salmon Production in the Northeast Pacific Ecosystem: Evidence of a Nutrient Deficit in the Freshwater Systems of the Pacific Northwest." *Fisheries* 25(1): 15–21.

Lackey, R. T. 2002. "Salmon Recovery: Learning from Successes and Failures." *Northwest Science* 76: 356–60.

Leopold, A. 1949. *A Sand County Almanac*. New York: Oxford University Press.

Lichatowich, J. A., G. R. Rahr, S. M. Whidden, and C. R. Steward. 2000. "Sanctuaries for Pacific Salmon." In E. E. Knudsen, C. R. Steward, D. D. MacDonald, J. E. Williams, and D. W. Reiser, eds., *Sustainable Fisheries Management: Pacific Salmon*, pp. 675–86. Boca Raton, Fla.: CRC Press.

Martin, G. 2003. "This Tasty Fish Didn't Get Away: Much-Sought Salmon Brought Back from the Brink by Efforts of Government, Farmers, Environmentalists." *San Francisco Chronicle*, January 31, p. A3.

Neuberger, R. L. 1941. "The Great Salmon Mystery: Will the Columbia River Developments Spoil the Salmon Runs?" *Saturday Evening Post*, 13 September 1941, pp. 20–21, 39–44.

Pollock, C. R. 1932. *Fortieth and Forty-first Annual Reports of the State Department of Fisheries and Game, Division of Fisheries*. Olympia, Wash.: State of Washington, Department of Fisheries and Game.

Rich, W. H. 1940. The Future of the Columbia River Salmon Fisheries. *Stanford Ichthyological Bulletin* 2: 37–47.

Stone, L. 1892. "A National Salmon Park." *Transactions of the American Fisheries Society* 21: 149–62.

"String of Supersize Salmon Sets Records." 2002. *Seattle Times*, Monday, December 23, p. B3.

FIGURE SOURCES

"Jumping Salmon" at start of book, "Salmon Family Tree," "Capital Salmon" at end of book, and chapter opening salmon art all courtesy of Ray Troll (© Ray Troll 2003).

CHAPTER 1: HISTORY, THE FIFTH H

Page 4   A Columbia River chinook salmon. Postcard mailed in September 1938.

CHAPTER 2: SALMON COUNTRY

Page 9   Seattle, Washington Territory, 1870. From "The Mediterranean of the Pacific," *Harper's New Monthly Magazine*, September 1870, p. 491.

Page 11   Salmon Leaping a Falls on its Migratory Trip to Spawning Grounds. From W. L. Finley, "Salmon or Kilowatts: Columbia River Dams Threaten Great Natural Resource," *Nature Magazine* (August 1935), p. 107.

Page 17   Logjam in the Queets River, Olympic National Park, ca. 1994. Photograph from David R. Montgomery.

CHAPTER 3: MOUNTAINS OF SALMON

Page 23   Selected rivers and creeks of western Washington State.

Page 27   Salmon family tree. Courtesy of Ray Troll (© Ray Troll 2003).

CHAPTER 4: SALMON PEOPLE

Page 42   An engraving made in 1778 shows a Nootka woman of Nootka Sound, British Columbia, with woven hat and a cedarbark cape. Image courtesy of the University of Washington Libraries, Manuscripts, Special Collections, and University Archives Division, negative no. NA3915.

Page 53   Columbia River area Indians fish with spears at Celilo Falls, Oregon, ca. 1910. Image courtesy of the University of Washington Libraries, Manuscripts, Special Collections, and University Archives Division, negative no. NA745.

CHAPTER 5: OLD WORLD SALMON

Page 65   Major salmon rivers of Great Britain. Adapted from A. Netboy, *The Salmon: Their Fight for Survival* Boston: Houghton Mifflin, 1974, p. 44.

Page 71   Queen Anne's Act for the restoration of Thames River salmon; composite of cover and first page.

Page 76   A print dating from 1846 of an adult salmon and a salmon grilse. From W. Jardine, ed., *The Naturalist's Library*, volume 36, *Ichthyology* (London: Chatto & Windus), "British Fishes," part 1.

CHAPTER 6: NEW WORLD SALMON

Page 90   Major salmon rivers of New England. Adapted from A. Netboy, *The Salmon: Their Fight for Survival* (Boston: Houghton Mifflin, 1974), p. 175.

Page 110   Salmon fishermen in New England in the 1880s. From H. P. Wells, "Salmon Fishing," *Harper's New Monthly Magazine* July 1886, p. 244.

Page 112   Ocean migration routes of the Atlantic salmon. Adapted from A. Netboy, *The Salmon: Their Fight for Survival* (Boston: Houghton Mifflin, 1974), p. 22.

CHAPTER 7: WESTERN SALMON RUSH

Page 137   Surveying the haul at Meyer's Packing Company, Seattle, 1895. Image courtesy of the University of Washington Libraries, Manuscripts, Special Collections, and University Archives Division, negative no. UW13761.

Page 138   Salmon halfway to the rafters of a Pacific coast cannery, date unknown. Image courtesy of the University of Washington Libraries, Manuscripts, Special Collections, and University Archives Division, negative no. UW184.

CHAPTER 8: BETTER THAN NATURAL

Page 151   Squeezing milt from chinook male salmon to fertilize eggs at Big White Salmon Station in Washington State, ca. 1920. From H. O'Malley, "Artificial Propagation of the Salmons of the Pacific Coast," in *Report of the United States Commissioner of Fisheries for 1919*, Bureau of Fisheries Document no. 879 (Washington, D.C.: U.S. Dept. of Commerce, 1920), appendix 2, plate 8.

Page 162   Bonneville salmon hatchery, showing rearing ponds, ca. 1917. From J. W. Cobb, "Pacific Salmon Fisheries," in *Report of United States Commissioner of Fisheries for 1916*, Bureau of Fisheries Document no. 839 (Washington, D.C.: U.S. Dept. of Commerce, 1917), appendix 3, plate 27.

CHAPTER 9: POWER FOR THE PEOPLE

Page 180   Locations of dams on the Columbia and Snake rivers.

Page 183   "Too bad the poor fish can't do this!" From W. L. Finley, "Are Salmon Now Sold Down the River? What Is the Attitude of the Commissioner of Fisheries?" *Nature Magazine* (August 1936), p. 107.

Page 197   Graph showing annual salmon catches on the Columbia River, 1860 to 2000. Source: Washington Department of Fish and Wildlife; Oregon Department of Fish and Wildlife.

CHAPTER 10: RIVERS OF CHANGE

Page 219   In November 2001 a chum salmon tries to get across a flooded road to reenter the Skokomish River. From *The Seattle Times*, November 16, 2001. Photo by Harley Soltes, courtesy of *The Seattle Times*.

Page 221   New development, levees, and fields on the floodplain of a river near Seattle in the 1990s.

CHAPTER 11: THE SIXTH H

Page 231   Cover illustration for R. D. Hume's *Salmon of the Pacific Coast* (San Francisco: Schmidt Label & Lithograph Company, 1893).

INDEX